International Conference "Mathematical Modelling in Biomedicine" 2019

ITM Web of Conferences Volume 31 (2020)

Moscow, Russia
30 September – 4 October 2019

Editors:

V. Volpert
F. Syomin

ISBN: 978-1-7138-0830-5

Printed from e-media with permission by:

Curran Associates, Inc.
57 Morehouse Lane
Red Hook, NY 12571

Some format issues inherent in the e-media version may also appear in this print version.

For additional information, please contact EDP Sciences – Web of Conferences
at the address below.

EDP Sciences – Web of Conferences
17, Avenue du Hoggar
Parc d'Activité de Courtabœuf
BP 112
F-91944 Les Ulis Cedex A
France

Phone: +33 (0) 1 69 18 75 75
Fax: +33 (0) 1 69 28 84 91

contact-edps@webofconferences.org

Additional copies of this publication are available from:

Curran Associates, Inc.
57 Morehouse Lane
Red Hook, NY 12571 USA
Phone: 845-758-0400
Fax: 845-758-2633
Email: curran@proceedings.com
Web: www.proceedings.com

TABLE OF CONTENTS

Statement of Peer review

In submitting conference proceedings to ITM Web of Conferences, I certify to the Publisher that I adhere to the **Policy on Publishing Integrity** of the journal in order to safeguard good scientific practice in publishing.

1. All articles have been subjected to peer review administered by the proceedings editors.
2. Reviews have been conducted by expert referees, who have been requested to provide unbiased and constructive comments aimed, whenever possible, at improving the work.
3. Proceedings editors have taken all reasonable steps to ensure the quality of the materials they publish and their decision to accept or reject a paper for publication has been based only on the merits of the work and the relevance to the journal.

Title, date and place of the conference International Conference "Mathematical Modelling in Biomedicine", September 30 - October 4, 2019. Moscow - Russia.

Proceedings editor(s): Vitaly Volpert, Fyodor Syomin

Date and editor's signature

27.02.2020

ITM Web of Conferences **31**, 01001 (2020) https://doi.org/10.1051/itmconf/20203101001
Mathematical Modelling in Biomedicine 2019

Chronobiological data: advantages of telemetric monitoring and prospects of mathematical modeling

Mikhail L. Blagonravov[1,*], *Björn* Lemmer[2,**], *Anna A.* Bryk[1,***], *Vyacheslav A.* Goryachev[1,****], *Abdukayum A.* Karimov[1,†], and *Sergey M.* Chibisov[1,‡]

[1]Peoples' Friendship University of Russia (RUDN University)

[2]Medical Faculty Mannheim University of Heidelberg

Abstract. Chronobiology and chronomedicine is a special part of biomedical sciences studying rhythmical patterns in physiological and pathological processes. In order to analyse probability of some pathology and to make forecasts concerning possibility of some diseases based on the signs of rhythmicity disorders it is necessary to perform continuous monitoring of different physiological functions for a certain period of time. Since 1984 until now we have had an excellent opportunity of studying biological rhythms and their disorders in animals using the method of radio-telemetric monitoring. A huge amount of continuous data obtained in telemetric monitoring could be used for mathematical modeling of different pathological processes on the basis of rhythmic patterns. In this work we have presented some preliminary results of the chronobiological study in which the effects of bright light on blood pressure and heart rate were investigated. The experiment was carried out on male rats of genetic strains: Wistar-Kyoto – normotensive rats and SHR – spontaneously hypertensive rats. The animals were exposed to 1 hour exposure of ~ 10000 lux white LED light from 10.00 to 11.00 a.m. For the analysis of daily profiles of blood pressure and heart rate we used the method of radio-telemetric monitoring of blood pressure and heart rate. It was shown that systolic blood pressure significantly increased in both Wistar-Kyoto and SHR rats under the action of bright during the time of bright light exposure (from 10.00 to 11.00 a.m.) and within the whole daytime period. For SHR rats an increase in diastolic blood pressure during the period of bright light action was also typical. But there were no significant changes in heart rate in the animals of either strain. These results require further and more detailed chronobiological studies to provide additional evidence. However traditional statistical methods seem to be important but not sufficient for further investigations. Moreover we could lose a considerable part of data without using contemporary methods of computer and mathematical modeling.

*e-mail: blagonravovm@mail.ru
**e-mail: bjoern.lemmer@medma.uni-heidelberg.de
***e-mail: bryckanna@gmail.com
****e-mail: goryachev-va@rudn.ru
†e-mail: abdukayum.karimov@bk.ru
‡e-mail: kalcna@mail.ru

1 Introduction

Chronobiology and chronomedicine as an independent part of Life sciences began to take shape relatively recently – in the 50-s of XX-th century, although the first mention of the cyclical nature of the onset and exacerbations of various diseases (primarily infectious) dates back to the time of Hippocrates.

In 1950s the concept of circadian (about-24-hour) rhythms - the main biological rhythms of living matter was formulated. In 1959 Franz Halberg (University of Minnesota) suggested the term *"circadian"* to describe about 24-hour rhythms [4]. In 1984 Jürgen Aschoff together with his colleagues from the United States, Britain, Germany, the Netherlands published a monograph "Biological rhythms". In this works J. Aschoff showed that it is light that stabilizes the circadian rhythms of the body [2].

Biological rhythms are a fundamental property of the living matter. Life without rhythm is impossible. Rhythmic activity is characteristic at different levels of organization of life: form unicellular to multicellular organisms. Along with the studies of normal biological rhythms, numerous investigations have shown that the pathogenesis of a number of diseases is closely related to the desynchronization of biological rhythms which is understood as loss of synchronization of internal and external rhythms which were previously synchronized.

2 Research methods used in chronobiology and chronomedicine

The principle of studying biological rhythms is different from the traditional approaches used in biomedical research. In this case it is necessary to use the monitoring technique of various body functions over a certain period of time (1 hour, 24 hours, several days, etc.). This makes it possible to assess the patterns of fluctuations of some parameters over time.

The methods such as ambulatory blood pressure monitoring (ABPM) and Holter monitoring of ECG were introduced into clinical practice long time ago. Using these methods, we obtain at least the results of 84 measurements of blood pressure (with ABPM) and a continuous ECG presentation (with Holter monitoring) within a 24-hour period. However disorders of biological rhythms commonly develop prior to the onset of some disease and may have a latent character.

In order to analyse probability of some pathology and to make forecasts concerning possibility of some diseases based on rhythmicity disorders it is necessary to perform continuous monitoring of different physiological functions for a given period of time which is sometimes impossible in clinical practice.

2.1 Radio-telemetry monitoring

1991 was an important milestone in the development of chronobiology, when the world's first wireless radio-telemetry monitoring technique was proposed [1]. From 1991 until now, we have had the opportunity to study biological rhythms and their disturbances in an animal experiment using the telemetry monitoring technique. The essence of this technique is that miniature devices are implanted in animals and measure blood pressure, heart rate, ECG, motor activity, body temperature and additional functions and to transmit information as a radio-signal to the receiving devices and storing data in the computer. Professor Björn Lemmer (Heidelberg University) was the first to introduce this method in experimental chronopharmacology in Europe [6].

The main advantages of telemetric monitoring in comparison with other research methods used to study the functions of organs and systems are as follows:

1. Continuous 24-hour monitoring of some parameters/functions for at least several months.
2. During the experiment animals are unrestrained and totally free moving. They do not suffer from pain, they are not stressed, they are not affected by anaesthesia.
3. We perform not just separate measurements of blood pressure, heart rate, body temperature and other indicators but continuous monitoring of these functions.
4. The experiments only need small number of animals to be valid.
5. Animals are not sacrificed due to the study.

The telemetry system consists of a radio transmitter (which is implanted into the animal's body during a surgical operation), a receiver (takes the radio-signal from transmitters) and a computer (saving data on the hard drive) [1, 6].

As a result, we get a huge array of data that can be analyzed using various statistical methods used in chronobiology and chronomedicine.

2.2 Programs and methods

In our investigations we commonly use the computer program "Chronos-Fit" which performs Rhythm Analysis as a combination of partial Fourier analysis and a stepwise regression technique. Rhythm analysis fits each harmonic separately and checks significance by F-test for each [7].

Other methods such as Cosinor Analysis or Wavelet Transform can serve as good tools but they also have some restrictions. At the same time a huge amount of continuous data obtained in telemetric monitoring could be used for mathematical modeling of different pathological processes on the basis of rhythmic patterns.

3 Some preliminary results

In the recent years attention is being increasingly focused on the problem of improper use (misuse) of artificial light. On the one hand the great majority of people all over the world lead their activity under artificial light patterns which are mainly characterized by a prolonged duration of daylight. Also "light pollution" which is characterized by excessive exposure of light in the night time is still a problem of great concern. Both these factors may induce the development of desynchrony – a complex of disorders caused by desynchronization of biological rhythms. On the other hand the aimed use of light is often successful when treating a number of mental and psychological disorders. In particular, bright light therapy (BLT) showed positive results with respect to seasonal affective disorders, non-seasonal depression, antepartum depression, sleep disorders, Parkinson's disease and so on. However, there is still little data concerning the effects of light therapy on functional activity of the cardiovascular system and particularly in case of hypertension. Many patients suffering from seasonal affective disorders, from different kinds of depression and so on can have some concomitant diseases, particularly hypertension of various genesis. We hypothesized that therapeutic effect of bright light may be accompanied by some changes in the biological rhythms of the cardiovascular system both in case of normal blood pressure and under hypertension.

We performed our experiments on male rats aged 32 – 34 weeks of 2 genetic strains: Wistar-Kyoto – normotensive rats and SHR – spontaneously hypertensive rats. The animals were kept in individual cages under an artificial light-dark schedule with light on at 7.00 a.m.

and off at 7.00 p.m., ~ 300 lux white light at eye level and nighttime absolute darkness. To estimate the effect of bright light the animals were exposed to 1 hour exposure of ~ 10000 lux white LED light from 10.00 to 11.00 a.m. For the analysis of daily profiles of blood pressure and heart rate we used the method of radio-telemetric monitoring which enables us to register the parameters continuously for 24 hours in totally freely moving animals. Radio-telemetry system (DSI, USA) [1] consists of a radio-transmitter (sensor) implanted in the animal body and monitoring blood pressure and ECG and sending data as a radio-signal to a receiver (nearby the cage), a computer for storing data. BP was monitored in the abdominal aorta. ECG was registered using the electrodes fixed to the animal chest along the electric axis of the heart. We began the experiments 10 days after implantation of the radio-transmitters. Blood pressure and ECG were monitored continuously for 24 hours the day before bright light treatment (control) and on the day of bright light treatment. The following parameters were estimated: mean daily (7.00 a.m. – 7.00 p.m.), mean daytime (7.00 a.m. – 7.00 p.m.) and mean nighttime (7.00 a.m. – 7.00 a.m.) values of systolic blood pressure and diastolic blood pressure.

Statistical data proceeding was carried out using Statistica 6.0 (StatSoft Inc.). The mean value and the error of the mean were calculated. The significance of difference in means was determined using the Mann-Whitney U test (the difference considered as significant at $p \leq 0.05$).

In the normotensive Wistar-Kyoto rats 1 hour exposure of bright light did not induce any significant changes in the values of blood pressure and heart rate for the 24-hour period. During the time of bright light exposure (from 10.00 to 11.00 a.m.) and within the whole daytime period systolic blood pressure was significantly higher in comparison with the corresponding periods of the previous day when bright light treatment was not performed (Figure 1). However the values of diastolic blood pressure or heart rate had no significant changes compared with the controls either at the time of bright light treatment or during the whole daytime period.

In the hypertensive SHR rats the values of both systolic and diastolic blood pressure were significantly higher during bright light treatment in comparison with the same period of the previous day when bright light treatment was not performed (Figure 2, left). The mean levels of daytime (10.00 a.m. – 7.00 p.m.) systolic (but not diastolic) blood pressure was also significantly increased compared with the controls which might indicate that the effect of bright light on systolic blood pressure remains notable even after the end of its exposure (Figure 2, right). For heart rate only a tendency to an increase was seen.

An increase in blood pressure under the action of bright light might seem quite expected as it was earlier shown that exposure to stressful stimuli contributed to aggravation of hypertension [3, 5]. However, according to our results, there was no significant elevation of heart rate either in normotensive Wistar-Kyoto or in hypertensive SHR rats which indicates the absence of true stress response. More possibly this is the activation reaction which may occur due to the stimulation of suprachiasmatic nuclei (SCN) of the hypothalamus. This hypothesis requires further study.

4 Conclusion

Our data demonstrate first that in chronobiological studies we do need additional statistical and mathematical modeling methods to evaluate and describe effects of drugs (many studies, see e.g. [8] or of bright light, as used in the present study.

Secondly, we could demonstrate the application of bright light in spontaneously-hypertensive rats, affected the level in blood pressure as well patterns in the 24-h profiles.

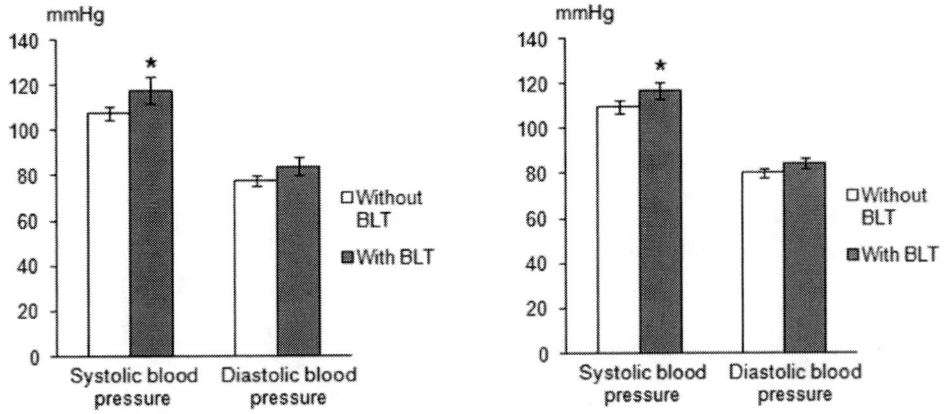

Figure 1. Left: Systolic and diastolic blood pressure in normotensive Wistar-Kyoto rats (from 10.00 to 11.00 a.m.). $^{*}p \leq 0.05$ blood pressure under bright light exposure (from 10.00 to 11.00 a.m.) vs the same period of time of the previous day when bright light was not exposed. Right: Systolic and diastolic blood pressure in normotensive Wistar-Kyoto rats during the daytime period (10.00 a.m. – 7.00 p.m.). $^{*}p \leq 0.05$ blood pressure on the day of light exposure vs the previous day when bright light was not exposed.

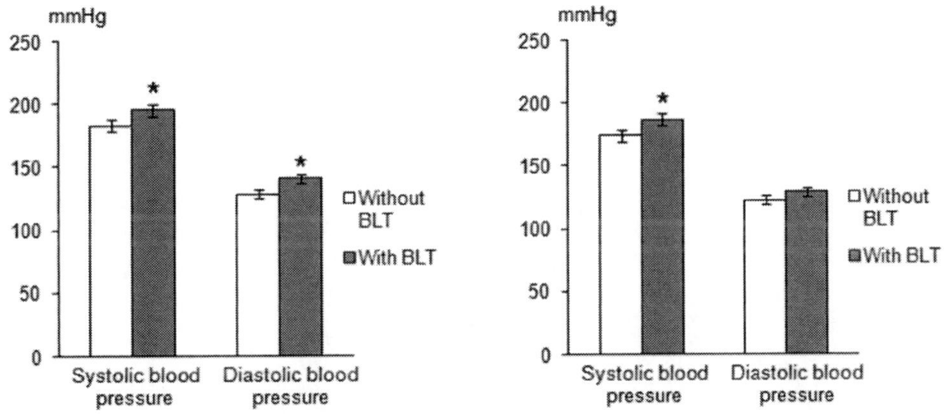

Figure 2. Left: Systolic and diastolic blood pressure in hypertensive SHR rats (from 10.00 to 11.00 a.m.). $^{*}p \leq 0.05$ blood pressure under bright light exposure (from 10.00 to 11.00 a.m.) vs the same period of time of the previous day when bright light was not exposed. Right:Systolic and diastolic blood pressure in hypertensive SHR rats during the daytime period (10.00 a.m. – 7.00 p.m.). $^{*}p \leq 0.05$ blood pressure on the day of light exposure vs the previous day when bright light was not exposed.

These results ask for further and more detailed chronobiological studies to provide additional evidence.

Performing our chronobiological investigations with the use of continuous 24-hour telemetric monitoring for many days we obtain a very big amount of data which could be used for the prognosis of the further development of various pathological conditions, in particular hypertension. By applying the traditional statistical methods to our results we can only assess the chronostructure of some physiological functions for within the period of monitoring.

https://doi.org/10.1051/itmconf/20203101001

Moreover we may miss a considerable part of such data without using contemporary methods of computer and mathematical modeling. So we could suggest that closer collaboration between chronobiology and computers sciences would give promising further results.

References

[1] Brockway B.P., Mills P.A., Azar S.H, Clin. Exp. Hypertens. **A 13**, 885-895 (1991)
[2] Eds. Aschoff, Jurgen, *Biological Rhythms* (Springer US, 1981) 563
[3] Folkow B., Hallbäck-Nordlander M., Martner J., Nordborg C., Acta Physiol. Scand., **116**, 133-139 (1982)
[4] Halberg F., Int. Z. Vitaminforsch Beih **10**, 225-296 (1959)
[5] Koepke J.P., Jones S., DiBona G.F., Brain Res., **404**, 80-88 (1987)
[6] Lemmer B., Mattes A., Bohm M., Ganten D., Hypertension, **22**, 97-101 (1993)
[7] Lemmer B., Gorbey S., Zuther P., *Chronos-Fit, personal edition, 2019*, Researchgate.net
[8] Lemmer B., Heart Failure Clinics, **13**, 739-757 (2017)

ITM Web of Conferences **31**, 01002 (2020) https://doi.org/10.1051/itmconf/20203101002
Mathematical Modelling in Biomedicine 2019

Shape effects in biological adhesion of ellipsoidal cells

Mikhail Kaznacheev[1,*] and *Aleksey* Belyaev[2,3,**]

[1]Department of Mathematics, Faculty of Physics, M.V.Lomonosov Moscow State University, Moscow 119991, Russia
[2]Department of Biophysics, Faculty of Physics, M.V.Lomonosov Moscow State University, Moscow 119991, Russia
[3]Interdisciplinary Research Center Mathematical Modelling in Biomedicine, RUDN University, 117198 Moscow, Russia

Abstract. Blood platelet adhesion is crucial for arterial thrombosis and hemostasis. The attachment of platelets to the injuries takes place under the action of high hydrodynamic forces and relies on the formation of breakable ligand-receptor bonds between the cell and the adhesive substrate. In this work we study how the geometrical effects may change the adhesive forces that stick platelets to the wounds. The mathematical model shows that oblate cells with high aspect ratio are more favourable for thrombus growth.

1 Introduction

Blood platelets play a key role in arterial and microvascular thrombosis, where the intensive hemodynamic flows and near-wall viscous stresses hinder the clotting of blood plasma [1, 2]. Physiological functions of platelets rely on their ability to adhere to a wound (or to a thrombus) and to withstand relatively mechanical stresses in the flowing blood [3]. In biological systems the adhesion mechanisms are mainly based on key-lock interactions of special protein molecules on the outer side their membranes with different adhesive substrates (e.g. collagen fibers, von Willebrand factor and membrane receptors of other cells) [1, 5–7]. The ligand-receptor bonds thus represent elastic bridges that emerge and dissociate with different kinetics, depending on the mechanical actuation and the molecular interactions of the involved proteins [8, 9]. A number of experimental techniques has been used to study the adhesive forces between platelets, most of which rely on mechanical separation of the adhering surfaces [10–12]. We concentrated our mind on geometric aspect of the problem. E.g. in work [13] authors investigated the influence of the form on collision effects, this required quite hard computational model. We tried to create a model which gives quantitative results in easier way. In this study, we hypothesised that the shape of the cell may cause a significant alteration of adhesive forces. We developed a mathematical model for the ligand-receptor adhesion of a blood platelet with a solid wall, covered by adhesive ligands. The model shows that the aspect ratio of the platelet is an important parameter, and that oblate cells are able to withstand greater pulling forces.

*e-mail: kaznacheev.michael@mail.ru
**e-mail: aleksey_belyaev@yahoo.com

2 Materials and Methods

2.1 Mathematical Model

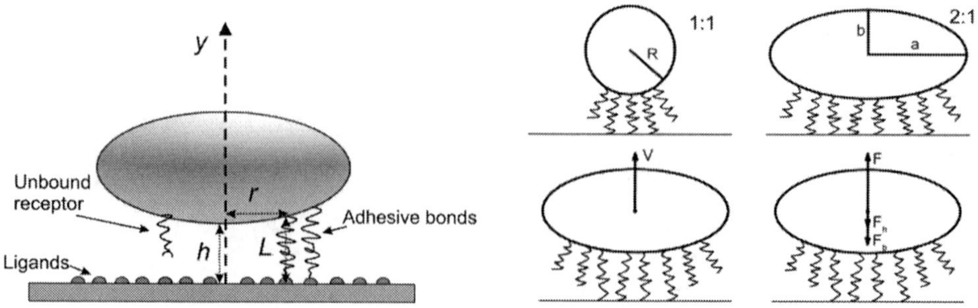

Figure 1. The scheme of the studied problem (left) and the illustration of the considered cases (right). The dynamics of adhesion of oblate ellipsoidal cell is considered for cells of different aspect ratios. The constant-velocity and constant-force regimes were studied.

We consider the biological adhesion of a platelet to a flat surface $y = 0$ evenly covered with adhesive ligand molecules (Fig. 1). We assume that the platelet surface is also uniformly covered with adhesion receptors. In addition, we may assume that only vertical bonds are formed. Then, the problem can be treated in a cylindrical system of coordinates (r,φ,y). In our model we consider the adhesion bonds to be springs with the equilibrium length λ and elasticity constant κ_0. The probability of bond formation is determined by the constants of forward and backward reaction rates between adhesion receptors and ligands. The platelet is approximated in this model by a rigid nondeformable ellipsoid of revolution. This is an admissible approximation, because in the deactivated state the platelet hardness is provided by the cytoskeleton [3], the platelet Young's modulus in this state is approximately 170 $pN \cdot \mu m^{-2}$[4]. Considering values of forces below it is sufficient to use such simplification. We consider the situation where the platelet moves only in vertical direction (normally to the wall). Different aspect ratios of the cell were considered in the presented study (Fig. 1).

The dynamics of the system relies on the reversible process of formation and dissociation of the adhesive bridges (or bonds) between the membrane receptors of the cell and the adhesive substrate (presented by immobilized protein ligands, e.g. collagen or von Willebrand factor [1]) on the wall. Since the number of adhesive receptors on the cell is typically very high (10^4-10^5), we describe this process in terms of bond density $n(r,t)$. The latter evolves in time according to the kinetic equation [8, 9]:

$$\frac{dn(r,t)}{dt} = K_{\mathrm{on}}(r,t)(N_r - n)(N_l - n) - K_{\mathrm{off}}(r,t)n(r,t), \tag{1}$$

$$K_{\mathrm{on}}(r,t) = K^0_{\mathrm{on}} \exp\left[\frac{-\kappa_s(L(r,t)-\lambda)^2}{2k_B T}\right], \tag{2}$$

$$K_{\mathrm{off}}(r,t) = K^0_{\mathrm{off}} \exp\left[\frac{(\kappa_0-\kappa_s)(L(r,t)-\lambda)^2}{2k_B T}\right], \tag{3}$$

where N_r is the density of receptors, N_l is the density of ligands, K^0_{off} and K^0_{on} are the constants of reaction rates in the equilibrium state, $L(r,t)$ is the bond length, k_B is the Boltzmann constant, T is the temperature, κ_0 and κ_s are the spring constants, and the sign in the second exponent determines the type of the bonds. Due to the axial symmetry of the system, there is no dependence on the azimuth angle φ in the equation. The bond length is given by

$$L(r,t) = h(t) - b\left(1 - \frac{r^2}{a^2}\right)^{1/2} + b, \tag{4}$$

where a and b are the ellipsoid semi-axes, h is the shortest distance between the ellipsoid and the surface at the initial time, r is the distance from the center of the system to the considered point, and t is the time.

The difference between different types of ligands (or/and receptors) can be taken into account by using different constants of reaction (i.d. different K^0_{off}, K^0_{on}). Below in Fig. 2(a) we represent the force kinetic with different value of $K_{eq} = \frac{K^0_{on}}{K^0_{off}} N_l$. This means changing the K^0_{on} or K^0_{off} (more in sec 3.1).

The force acting upon the platelet is computed also by integrating over all possible positions of bonds with account for the tension of bonds. Because we consider only vertical bonds, the force has the only component in the y-direction given by:

$$F_b(t) = \kappa_0 \int_0^a \int_0^{2\pi} n(r,t)[L(r,t) - \lambda] r dr d\varphi = 2\pi\kappa_0 \int_0^a n(r,t)[L(r,t) - \lambda] r dr. \tag{5}$$

In our model we investigate only vertical bonds, but on 2D case, thus we go to the polar coordinate system. This leads to necessity to integrate over a circle thus we have the factor r and integral over $drd\varphi$ in spite of nothing depends of φ. So, it gives the factor 2π in the final result. Hence in reality we have the 1D integral over r only.

In this study we modelled two different situations: pulling the platelet from the wall (i) with a constant vertical velocity V, and (ii) by a constant external force F directed vertically, Fig.1. These are typical setups for measuring interfacial forces by Surface Force Apparatus [12]. For the constant velocity regime we measured the adhesive force F_b, and the distance between the platelet and the wall was explicitly time-dependent:

$$h(t) = h_0 + V \cdot t. \tag{6}$$

In the case of pulling by the externally applied force F, we also should take into account the viscous hydrodynamic forces arising from the fluid motion around the platelet. Naturally, the Reynolds number for a platelet-sized particle in a fluid with the viscosity of blood plasma is of order of 0.1 and less. Therefore, in our model the hydrodynamic interaction is characterized by a lubrication force [12, 14], given by the following formula

$$F_h = \frac{6\pi\eta R^2_{\text{eff}} \frac{dy}{dt}}{y}, \quad R_{\text{eff}} = \frac{a^2}{b}, \tag{7}$$

where R_{eff} is the effective hydrodynamic radius, η is the dynamic viscosity, y is the distance between the cell membrane and the wall. On assuming the over-dumped dynamics of the cell at low Reynolds numbers, the equation for the motion of the platelet may be derived from the previous expression:

$$\frac{dy(t)}{dt} = \frac{1}{6\pi\eta R^2_{\text{eff}}} [F - F_b(t)] \cdot y(t), \tag{8}$$

where F is the external (pulling) force on the platelet and F_b is the tension force of adhesive bonds. The latter one is obtained by a summation over all attached (or active) bonds at each moment of time. Thus, we solve Eq.(1) for the bond density at a given moment of time, compute the force of adhesion interaction with formula (5), and then use Eq. (8) to find the platelet position in the next moment time.

Solving the kinetic equation can give quantitative information about influence of the certain aspect ratio on the adhesion process. Moreover, the influence of other parameters like receptor density, bond constants, force or velocity can be investigated. The kinetics of the adhesion process may also be derived from the mathematical model.

Table 1. Model parameters

Parameter	Symbol	Value	Dimension
Initial distance	h_0	0.1	μm
Length of relaxed bond	λ_0	0.1	μm
Bond formation rate	K^0_{on}	10^{-15}	$\mu\text{m}^2 \cdot \text{s}^{-1}$
Bond dissociation rate	K^0_{off}	10	s^{-1}
Spring constant	κ_0	10^{-5}	$\text{N}\cdot\text{m}^{-1}$
Spring constant	κ_s	$0.5 \cdot 10^{-5}$	$\text{N}\cdot\text{m}^{-1}$
Thermal energy	k_bT	$4,14 \cdot 10^{-21}$	J
Density of receptors	N_r	10^4	μm^{-2}
Density of ligands	N_l	10^4	μm^{-2}

2.2 Numerical Methods

To solve the differential equation we use the one stage Rosenbrock scheme with a complex coefficient [15]. We controlled the accuracy using the recurrent Richardson method. We consider the Cauchy problem

$$y' = f(y,t), \quad y(0) = y_0, \tag{9}$$

where y is the sought function and $f(y,t)$ is the given function. For this problem, the CROS1 scheme has the form

$$\left[1 - \tau\alpha f_y(y_n, t_n)\right] w_n = f\left(y_n, t_n + \frac{\tau}{2}\right), \tag{10}$$

$$y_{n+1} = y_n + \tau Re(w_n), \tag{11}$$

where $\alpha = \frac{1+i}{2}$, and τ is the time step. For each r the bond kinetics Eq. (1) is an ordinary differential equation solvable with the numerical method Eqs. (10) and (11). We ensure the convergence of the numerical scheme by Richardson's procedure, which is based on mesh refinement [15]:

$$R_{kN} = \frac{y_{rN} - y_N}{r^p - 1}, \quad p_k = \frac{\lg \frac{R_{k-1}}{R_k}}{\lg r}, \tag{12}$$

where k is the index that determines the number of the mesh (0 is the initial mesh, 1 is the mesh refined once, and so on), r is the factor of mesh refinement (in what fold the mesh is refined in comparison with the previous one), R_{kN} is an inaccuracy of the numerical solution, y_N is the numerical solution to the equation, y_{rN} is the solution on a refined mesh, and $p = 2$ is the accuracy of the numerical scheme. Here p_k represents the *a posteriori* accuracy of the numerical scheme, calculated according to Eq.(12) with several different meshes. If p_k converges to the theoretical value $p = 2$ upon increase of the mesh density, then the numerical scheme works correctly.

3 Results

3.1 Pulling with a constant velocity

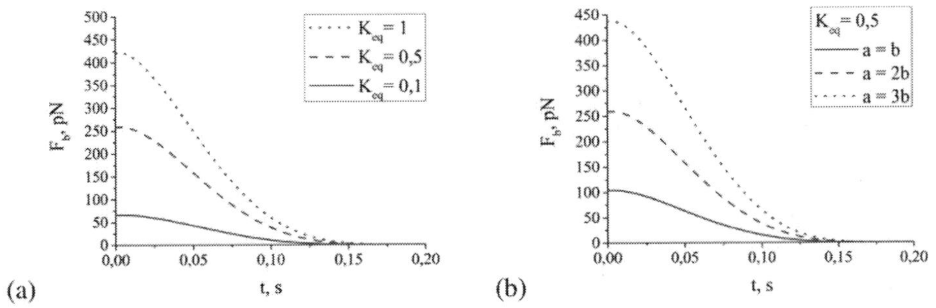

Figure 2. The adhesive force F_b over time in the constant velocity case for different K_{on} (left) and various aspect ratios of the platelet (right).

Firstly we investigated the dependence of the adhesive force on the binding kinetics (Fig. 2(a)). We altered the equilibrium constant for the binding-dissociation reactions $K_{\text{eq}} = K^0_{\text{on}}/K^0_{\text{off}}$. When K_{eq} becomes smaller, the force F_b decreases as well.

We studied the dependence of the simulation results on the shape of a platelet Fig.2(b). The semi-axes were chosen in the following manner. For the reference, we took the aspect ratio 2 : 1 ($a = 1.5$ μm and $b = 0.75$ μm) which is a typical value for human blood platelets. We altered the aspect ratio, keeping the platelet volume (7.07 μm^3) constant. The velocity of the platelet motion in these calculations was set to $0.4\ \mu m \cdot s^{-1}$. As the ratio of semi-axes $a : b$ increases, the adhesive force also increases, which is a consequence of the larger area on which the bonds may be formed. Thus, for the aspect ratio 3 : 1, the force of adhesion appears to be more than twice that for 1 : 1.

Other way to increase the force is to increase the receptors density. Increasing the number of receptor density and the flattening a platelet can work together in increasing the ability of adhesion. We suppose that these mechanisms have important role both in vivo. The increase of the receptor density and the flattening of the platelet can work together enhancing the adhesion. But they have different impact on the number of bonds. Fig.3(a) shows how the number of bonds depends on the aspect ratio, here the density of receptors is $10^4 \mu m^{-2}$. The next picture (Fig.3(b)) represents how the kinetics depends on different numbers of receptors for a given aspect ratio 2 : 1. We can see that the number of bonds can significantly change if the receptor density is altered by the order of amplitude or more. The alteration of the aspect ratio also leads to a fast change of the amount of bonds. Therefore, flattening of the shape is a an efficient way to attain strong adhesion, especially when the receptor density is already very high and/or can not be increased easily. We also note that the aspect ratio governs the distribution of the bonds over the platelet membrane (Fig.3(c)) and, consequently, the distribution of the mechanical tension arising from the bonds.

3.2 Pulling by a constant force

The width of the minimum gap between the platelet and the plane has the form depicted in Fig.4 at the constant force $F = 200$ pN. At the semi-axes ratio 3 : 1 the larger number of bonds are generated than at the semi-axes ratio 1 : 1. Therefore, a spherical cell (1 : 1) is

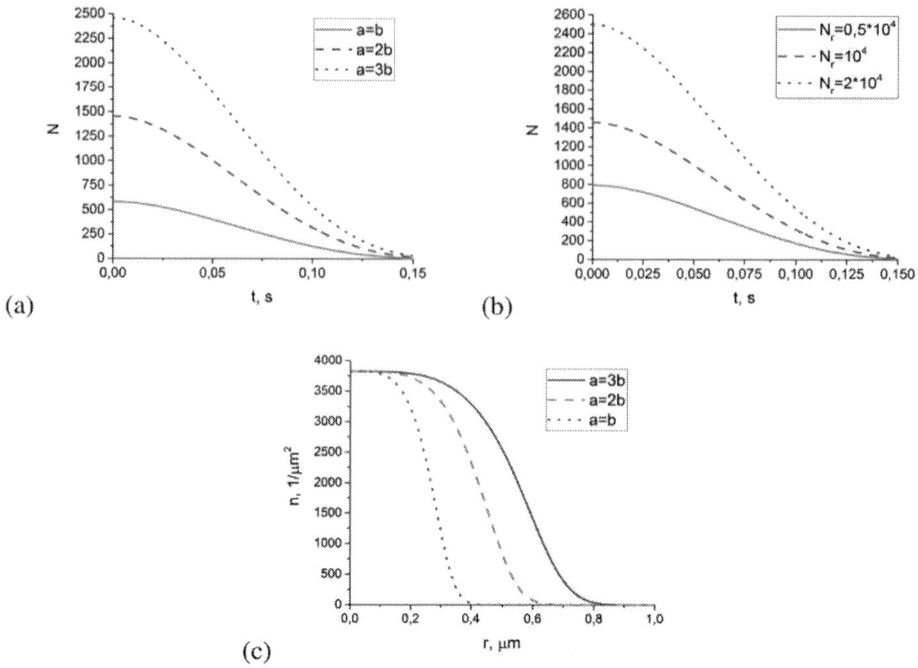

Figure 3. (a) Number of bonds N over time in the constant velocity case for different aspect ratio. (b) Number of bonds N over time in the constant velocity case for different receptor densities N_r. (c) Bond density n over the distance from the center of the cell r.

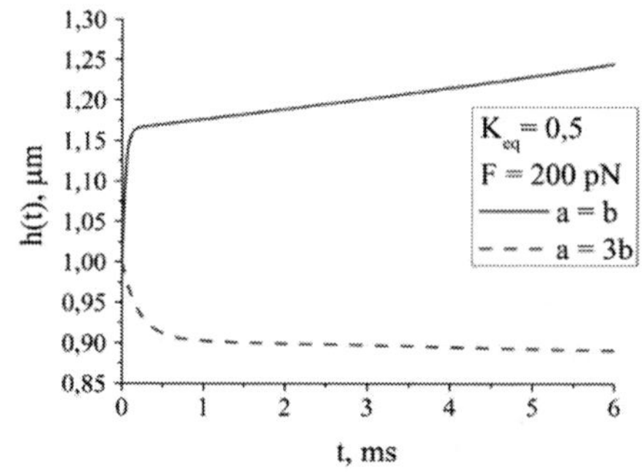

Figure 4. Width of the minimum gap between the platelet and plane over time, F = 200 pN.

not able to form the required amount of bonds which gives sufficient value of force to stay the platelet near the wall. The spherical cell is detached from the surface, being pulled by the external force to the bulk of the fluid, Fig 4. As the semi-axes ratio increases, the shape of the cell becomes more plate-like, and the effective area of the platelet is larger and, as a consequence, a larger number of bonds is formed, hence if the F_b is high enough, it balances the external force F, and the platelet rests on the wall, Fig 4. Shape is thus an important factor for platelet adhesiveness, which may have important implications for the mechanics and rheology of platelet thrombi. Oblate spheroidal shape favours the bond formation, so that platelets with high aspect ratio can withstand greater pulling forces.

3.3 Discussion

The results provide a numerical basis for the interpretation of measurements of biological adhesion forces, depending on the cell's shape, as well as the density of cell adhesive molecules. We considered the adhesive kinetics based on Bell's law of bond dissociation [7]. However, in several cases a more complex behavior has been observed [17].

For further studies it is instructive to consider the vWF and GP1b with the dynamics of catch- and slip-bonds observed in atomic force microscopy [16] and flow chamber experiments [17]. Another possible improvement of the model is to simulate the platelet motion under hydrodynamic flows, typical for flow chamber experiments and studies *in vivo*. In this case the interplay between hydrodynamic forces, cell-cell collisions and bonds kinetics may lead to complex dynamics of cells and their aggregates. Finally, since platelets have several adhesive receptors, activated by force or chemical agonists, it would be interesting to simulate temporal dynamics of adherent cells after their activation. In this case, the elasticity of platelets also undergoes dramatic changes followed by to the cytoskeleton restructuring.

This study was supported by Russian Foundation for Basic Research (RFBR grants 19-02-00480-a , 19-31-70002-mol-a-mos). AVB also acknowledges the support from RUDN University (5-100 project).

References

[1] Jackson S.P., Nesbitt W.S. and Westein E., Journal of Thrombosis and Haemostasis **7(Suppl. 1)**, 17–20 (2009).
[2] Belyaev A.V., Dunster J.L., Gibbins J.M., et al., Phys Life Rev **26-27**, 57 (2018).
[3] Michelson A.D., *Platelets 3rd Edition* (Academic Press, 2013).
[4] Haga et al., Annals of Biomedical Engineering **26**, 268–277 (1998).
[5] Kulkarni S.,Dopheide S.M., Yap C.L., et al., J Clin Invest **105**, 783 (2000).
[6] Moore N.W., Mulder D.J., Kuhl T.L., Langmuir **24**, 1212 (2008).
[7] Bell G.I., Science **200(4342)**, 618 (1978).
[8] Hammer D.A., Apte S.M., Biophys J **63**, 35 (1992).
[9] Sircar S., Bortz D. M., Math Biosci **245(2)**, 314 (2013).
[10] Rakshit S., Zhang Y., Manibog K., et al., PNAS **109(46)**, 18815 (2012).
[11] Arya M., Kolomeisky A.B., Romo G.M., et al., Biophys J **88**, 4391 (2005).
[12] Israelachvili J., Min Y., Akbulut M., et al., Rep Prog Phys **73(3)**, 036601 (2010).
[13] Mody and King, Biophys J **95**, 2539–2555 (2008).
[14] Asmolov E.S., Belyaev A.V., Vinogradova O.I., Phys Rev E **84**, 026330 (2011).
[15] N.N. Kalitkin, P.V. Koryakin, *Numerical methods, book 2* (2013) 304.
[16] Lining Ju et al., Thrombosis Research **136**, 606–612 (2015).
[17] L.A.Coburn et al., Biophys J **100**, 304–312 (2011).

ITM Web of Conferences **31**, 01003 (2020) https://doi.org/10.1051/itmconf/20203101003
Mathematical Modelling in Biomedicine 2019

Simulation of lymph flow with consideration of natural gravity force influence

Anastasiia Mozokhina[1,*] and *Gennadii* Lobov[2,**]

[1]Lomonosov Moscow State University, 119991, Moscow, Russia
[2]Pavlov Institute of Physiology, Russian Academy of Sciences, St. Petersburg 199034, Russia

Abstract. The model of lymph flow in the human lymphatic system in the quasi-one-dimensional approach is considered in the paper. Calculations with natural gravity force influence are presented. Calculations show that in some sets of parameters of lymphatic vessels the physiological flow can be reached in the model with zero global pressure gradient influence only by the "muscle" pump.

1 Introduction

The lymphatic system is a vessel system in the human body. It performs many functions, in particular, transport, drainage, immune, and depositing ones. In considering the transport function, LS consists of lymphatic vessels and lymph nodes. Lymphatic vessels have a specific structure, which provides the unidirectional flow of lymph: from the interstitial space to the upper vena cava. According to available scanty physiological observations, lymph flow in the LS is influenced by the pressure gradient, active and passive contractions of lymphatic vessels. Pressure in the interstitial space is close to zero, as well as pressure in the upper vena cava, so the global pressure gradient is small, can even be zero or negative. The pressure of lymph in the lymphatic vessels essentially is unknown. According to some sources, it can reach 20-40 mm Hg in the thoracic duct [1]. Lymphatic vessels are divided by valves into segments which are called lymphangions. These lymphangions can produce active contractions. Active contractions with the prevention of backward flow of lymph by valves form a "muscle" pump, which is one more force propagating lymph in the LS. Active contractions have periodic - "phase" - part and tonic part, which is a reaction of the lymphatic vessel wall to the pressure difference in the lumen. Passive contractions are changes in the vessel lumen as a result of compression from surrounding tissues and the massage influence of big blood vessels. The degree of influence of each mechanism affecting lymph flow is not clear now, and we hope that current work can bring us closer to the understanding of mechanisms lying under lymph propagation in the human LS. Such understanding, in turn, can spread some light to the reasons for diseases related to the LS, for example, lymphedema.

2 Model of lymph flow

To get a model of lymph flow in the human lymphatic system in the quasi-one-dimensional approach, we created a graph of human LS. This graph is anatomically adequate, spatially

*e-mail: asm@cs.msu.ru
**e-mail: gilobov@yandex.ru

oriented, consistent with a similar graph of the cardiovascular system [2], and is suitable for numerical calculations. The basis of the graph is regional groups of lymph nodes, which connected according to data from [3]. Arcs of the graph represent lymphatic vessels and lymph nodes, and vertexes of the graph represent points of vessel bifurcations, rare valves, and boundaries in the interstitial space and upper vena cava. All arcs of the graph are divided into four groups depending on mechanisms influencing lymph flow in the vessels of the group. These are big vessels with contractions, collectors with contractions and frequent valves, lymph nodes with contractions, and lymphatic capillaries, which are modeled by effective representations with no contractions and valves [4].

Lymph flow with no respect to contractions and frequent valves and respect to normal gravity force influence is modeled by a system of quasi-one-dimensional hemodynamic equations [2, 5]:

$$\frac{\partial s}{\partial t} + \frac{\partial us}{\partial x} = 0, \frac{\partial u}{\partial t} + u\frac{\partial u}{\partial x} + \frac{1}{\rho}\frac{\partial p}{\partial x} = -8\pi\nu\frac{u}{s} + g\cos\varphi, s = s(p) = s_0 + \theta(p - p_0), \quad (1)$$

where x is a spatial coordinate, t it time, unknown functions are velocity $u(x,t)$ and pressure $p(x,t)$, $s(p)$ is a cross-sectional area which is a linear function of pressure p, ρ is density, ν is viscosity, g is a gravity factor, φ is an angle between spatial axis of the vessel and gravity force vector. The third equation is so-called "state" equation, it contains all physiology. s_0 and p_0 are parameters of the vessel, θ describes elasticity of the vessel: the more θ is, the more elastic vessel is.

Flow in the vessel with frequent valves are modeled by changes in the right part of the second equation of (1):

$$\frac{\partial u}{\partial t} + u\frac{\partial u}{\partial x} + \frac{1}{\rho}\frac{\partial p}{\partial x} = -8\pi\nu(u)\frac{u}{s} + g\cos\varphi, \nu(u) = \frac{\nu_{max} - \nu_{min}}{\pi}\operatorname{arccot}(Cu + D) + \nu_{min}. \quad (2)$$

Here ν is a function of velocity u which gives normal value of viscosity ν_{min} when lymph flows along with valves, and much more value ν_{max} is the case of backward flow [6, 7].

Influence of contractions of the vessel are taken into account in the third equation of (1)

$$s = s_0 + \theta(p - p_0) + \theta A\sin\left(\frac{2\pi}{\lambda}(x - at)\right). \quad (3)$$

The third term is the periodic part of active contractions. This periodic part is modeled by a sine function, which is simple enough, still can give rich flow patterns, and so is widely used for such purposes [8–10]. Parameters are amplitude of the contraction A, wavelength λ, and velocity a. We also use frequency $\omega = 2\pi a/\lambda$ to describe contractions.

The graph is created and the following calculations are performed in the Cardio Vascular Simulation System (CVSS) [2]. To solve the nonlinear system (1) and its variations with equations (2) or (3) the finite difference schema with first-order time and second-order space approximation is used [2, 11–14].

3 Calculations with natural gravity force influence

3.1 Test model

The test model consists of one vessel placed in the gravity force. The vessel is vertically oriented in the way that gravity force influences in the direction against valves. We take the length of such vessel 200 cm, so this is the simplest model of LS. Let us estimate pressure gradient needed to force lymph to flow against gravity force: $dp = \rho gl =$

Table 1. Results of calculations of the test model

No	description	0	70	120	130	140	150	160
1	no valves				-0.013		0	0.006
2	$v_{max} = 4$		-0.0005	0	0			
3	valves type 1	0			0		0	

$1\,g/cm^3 \cdot 1000\,cm/s^2 \cdot 200\,cm \approx 150\,mm\,Hg$. So if we want lymph to flow against gravity force, the pressure gradient must be more then 150 mm Hg. Results of calculations (table 1, No 1) are consistent with this estimation, a positive value of flux is flow in the direction from interstitial space to upper vena cava against gravity force, a negative value is flow along with gravity force.

Other calculations in one vessel show that global dp can be reduced by valves. There are two types of valves in the model: valves in the big lymphatic vessels which lymphangions more than 1 cm (the first type, modeled by a condition of zero backward flux), and valves in vessels with lymphangions of 1-2 mm (the second type). With valves of the second type presented dp can be reduced to 120 mm Hg (table 1, No 2), and it can be reduced to 0 mm Hg in the case when valves of the first type are presented (table 1, No 3). Let us consider a model with the valve of the first type in the middle of the vessel. In this case, the pressure under the valve is -74.2 mm Hg and above the valve $p = 76.2$ mm Hg, so the pressure difference is 150.4 mm Hg in the model, but now pressure in interstitial space and upper vena cava have physiological values 1 mm Hg.

As shown by calculations in the test model, valves can stop lymph from flowing down along with the gravity force, but cannot give a positive nonzero flux. Let us try to get a positive value of flux by adding contractions in the model.

Calculations are performed in one vessel with different length l. There are valves of type 2 with $v_{max} = 4\,cm^2/s$ and contractions with $A = 0.1$ mm Hg, $\lambda = 1$ cm, and $a = 1.5$ cm/s ($\omega = 3\pi$), pressure is 1 mm Hg in the boundaries. For $l = 1$ cm, $l = 10$ cm and $l = 100$ cm time-average output flux is the same $y = -0.00035$ ml/s. These results show that length does not influence output flux, so further calculations are performed in the vessel of length 1 cm.

The negative value of flux (time-average integral) shows, that lymph flows along with gravity force against valves. The positive value of flux can be got by alerting parameters of the vessel. Results of calculations with different parameters of the vessel and increasing efficiency of the "muscle" pump [15] are shown in table 2. By changing parameters of the vessel it is possible to get positive flux with $g = 1000\,cm/s^2$ (table 2, No 6-8, 11-16). The increase of g leads to flux decrease. The increase of amplitude A leads to the increase of output flux when parameters of the vessel stay the same (table 2, No 1, 6, 11-13, No 14-16). Pressure gradient appearing in the vessels as a result of calculations less than 1 mm Hg. These calculations show that valves with contractions don't only stop lymph from flowing along with gravity force, but also give a nonzero output flux to the upper vena cava.

3.2 Calculations on the graph

Now we perform calculations on the LS graph (figure 1). Parameters of vessels are shown in table 3, $p_0 = 0$ mm Hg for all arcs, θ from equation (3) has values θ_1 from table 3. Pressure in interstitial space is 3.4 mm Hg. Condition of free pressure is stated in the upper vena cava: $p' = 0$. There are valves of second type with $v_{max} = 4\,cm^2/s$, contractions with $A = 0.1$ mm Hg, $\lambda = 1$ cm, $a = 1.5$ cm/s ($\omega = 3\pi$) and 25 valves of the first type in the model. Calculations in the horizontal position ($g = 0$) and vertical position ($g = 1000\,cm/s^2$) give us time-average integral fluxes 0.002 ml/s and 0.33 ml/s respectively. Flux in vertical case much more, then

Table 2. Calculations of test model with different $\theta\,(cm^2/mm\,Hg)$ and contraction parameters, 0 - 1000 are values of g in cm/s^2. There are values of time-average integral flux in the cells under values of g, $10^{-5}\,ml/s$. A (mm Hg) — amplitude of contractions, Δs (%) — difference in cross-section area, dp (mm Hg) — pressure gradient, appearing in the system as a result of calculations

No	A	Δs	θ	0	400	500	600	700	1000	dp
1	0.1	1.83	0.00165	7.8	1.1	0.3	-0.4		-1.4	0.17
2	0.1	1.81	0.00163	8		0.2	-0.5		-1.5	0.23
3	0.1	0.34	0.00031	1		0.1	0.02	-0.005	-0.08	0.17
4	0.1	2.48	0.00223	13	0.05	-1	-3		-5	0.19
5	0.1	2.65	0.00238	13	1	-0.2			-3	0.2
6	0.2	3.67	0.00165	15					0.4	0.37
7	0.2	3.63	0.00163	16					0.2	0.42
8	0.2	0.68	0.00031	2					0.2	0.38
9	0.2	5.3	0.00238	25				3	-0.6	0.26
10	0.2	4.96	0.00223	26			3	2	-3	0.36
11	0.5	9.17	0.00165	32					14	0.56
12	0.9	16.5	0.00165	55					31	0.63
13	1	18.3	0.00165	59					35	0.73
14	5	27.5	0.0005	150					120	0.7
15	9	49.5	0.0005	250					230	0.86
16	10	55	0.0005	280					250	0.81

flux in horizontal case, it happens because there is no control for total lymph volume in the model. Flux from head and neck grow with increase of g and output flux grows as well.

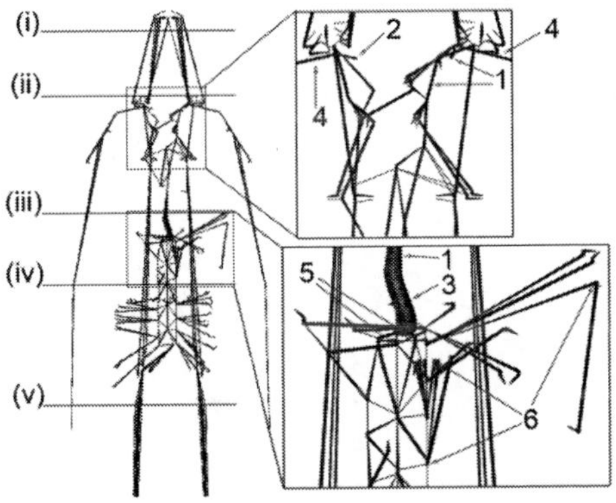

Figure 1. The graph of the lymphatic system: (i) — head, (ii) — neck, (iii) — diaphragm, (iv) — elbows, (v) — groin; 1 — thoracic duct, 2 — right lymphatic duct, 3 — cisterna chyli, 4 — subclavian trunks, 5 — lumbar trunks, 6 — lymph nodes.

In the next set of calculations, we control input flux from the head and neck. Now the condition of constant input flux $q = 1 \cdot 10^{-5}$ ml/s is stated in the inputs in the interstitial fluid of head and neck. There are $p = 10$ mm Hg in other inputs, and $p = 10$ mm Hg in upper vena cava. Valves of both types are included in the model, $v_{max} = 4$. Contractions have following parameters: $\lambda = 1$ cm, $a = 3$ cm/s ($\omega = 6\pi$). Parameters of the vessels are listed in table 3, θ from equation (3) has values θ_2 from table 3. The results of the calculations are presented in figure 2 (circles). The increase in the amplitude of contractions A leads to an increase in

Table 3. Parameters of vessels of the graph

Vessel	s_0, cm^2	θ_1, $cm^2/mmHg$	θ_2, $cm^2/mmHg$
Cisterna chyli	0.0503	0.02495	0.00468
Thoracic duct	0.0126	0.00623	0.00117
Right lymphatic duct	0.00316	0.00158	0.0003
Lumbar trunks	0.00708	0.00354	0.00066
Other trunks	0.00316	0.00158	0.0003
Collectors	0.0036	0.0015	0.00028
Effective	0.002	0.001	0.00019
Lymph nodes	0.0126	0.16457	0.03096

output flux. The physiological value of flux 0.0352 ml/s ≈ 3 l/day is obtained with amplitude $A = 3$ mm Hg which is 34% difference in diameter ($\Delta s = 11.3\%$). The same quality results are obtained for the pressure 5 mm Hg (figure 2, squares) and 3 mm Hg (figure 2, triangles) in the interstitial space and upper vena cava. Decreasing of pressure in the interstitial space and upper vena cava (characteristic pressure in the calculation) leads to increasing of flux in the upper vena cava. The increase of pressure difference (by decreasing pressure in upper vena cava) also leads to an increase in output flux. Pressure gradient appearing as a result of calculations is shown in table 4, it is 20-30 mm Hg and maximum pressure reached in the calculations is about 11-25 mm Hg (table 4 in parentheses).

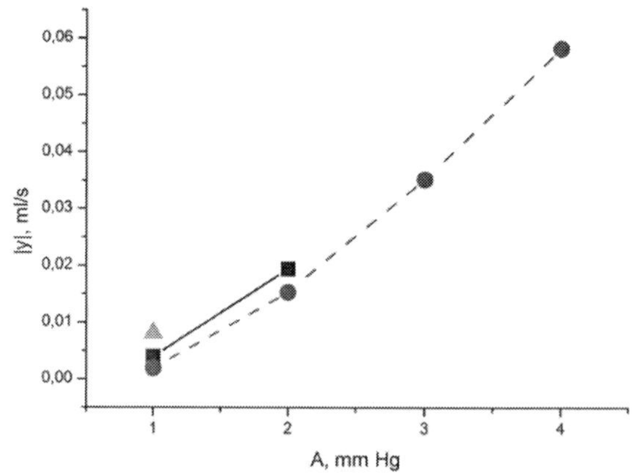

Figure 2. Results of calculations with natural gravity force influence with different characteristic pressures (in the interstitial space and upper vena cava, zero pressure gradient) and amplitudes of contractions. A — amplitude of contractions, $|y|$ — average integral flux to the upper vena cava; squares — characteristic pressure is 5 mm Hg, circles — 10 mm Hg, triangles — 3 mm Hg

4 Conclusion

Calculations on the graph show that the "muscle" pump (active contractions of the vessel with valves) can give a physiological value of output flux by itself in zero global pressure

Table 4. Pressure appearing as a result of calculations in the LS graph with natural gravity force influence. A (mm Hg) — amplitude of contractions, p (mm Hg) — characteristic pressure in the calculation (initial pressure gradient is zero). There is resultant maximum pressure gradient (mm Hg) in the cells and values of maximum pressure (mm Hg) in the parentheses

A	$p = 10$	$p = 5$	$p = 3$
1	25.7 (17.19)	20.27 (11.19)	25.5 (14.7)
2	26.27 (18.44)	21.67 (13.01)	
3	27.15 (22.73)		
4	27.88 (25.26)		

gradient. It means that active contractions can be the main force propagating lymph in the LS at least hydrodynamically.

References

[1] Petrenko V.M., *Functional morphology of lymphatic vessels* (DEAN, St. Petersburg, 2008) p. 400 (in Russian)
[2] Abakumov M.V., Gavrilyuk K.V., Esikova N.B., Lukshin A.V., Mukhin S.I., Sosnin N.V., Tishkin V.F., Favorskii A.P., Differential Equations **33**, 895-901 (1997)
[3] Borzyak E.I., Bocharov V.Ya., Sapin M.R., *Human anatomy* (Medicina, Moscow, 1993) p. 560 (in Russian)
[4] Mozokhina A.S. and Mukhin S.I., *Trends in Biomathematics: Modeling, Optimization and Computational Problems: Selected works from the BIOMAT Consortium Lectures* (Springer, Cham, 2018) 325-334.
[5] Bunicheva A.Ya, Menyailova M.A., Mukhin S.I., Sosnin N.V., Favorskii A.P., Mathematical Models and Computer Simulations **5**, 81-91 (2013)
[6] Simakov S., Gamilov T., Soe Y.N., Russian Journal of Numerical Analysis and Mathematical Modelling **28**, 485-504 (2013)
[7] Mozokhina A.S., Mukhin S.I., Differential Equations **54**, 938-944 (2018)
[8] Macdonald A.J., Arkill K.P., Tabor G.R., McHale N.G., Winlove C.P., Am. J. Physiol. Heart. Circ. Physiol. **295**, 305-313 (2008)
[9] Bertram C.D., Macaskill C., Moore Jr.J.E., J. Biomech. Eng. **133**, (2011)
[10] Rahbar E., Moore J.E.Jr., J. Biomech. **44**, 1001-1007 (2011)
[11] Ashmetkov I.V., Khrulenko A.B., Mukhin S.I., Sosnin N.V., Favorskii A.P., Differential Equations **36**, 1021-1026 (2000)
[12] Abakumov M.V., Esikova N.B., Mukhin S.I., Sosnin N.V., Tishkin V.F., Favorskii, A.P. *Finite difference scheme for the solution of hemodynamic problems on the graph* (Dialog-MSU, Moscow, 1998) p.16 (in Russian)
[13] Ashmetkov I.V., Mukhin S.I., Sosnin N.V., Favorskii A.P., Khrulenko A.B., *Numerical investigation of properties of the finite difference scheme for hemodynamic equations* (Dialog-MSU, Moscow, 1999) p.14 (in Russian)
[14] Mozokhina A.S., Mukhin S.I., Mathematical Models and Computer Simulations **11**, 894–904 (2019)
[15] Mozokhina A.S., Mukhin S.I., Lobov G.I., Russ. J. Numer. Anal. Math. Modelling **34**, 261–268 (2019)

Electrophysiological Biomarkers for Age-Related Changes in Human Atrial Cardiomyocytes: In Silico Study

Tatyana Nesterova[1,2,*], *Konstantin* Ushenin[1,2,**], *Dmitry* Shmarko[1,2], and *Olga* Solovyova[1,2,***]

[1]Institute of Immunology and Physiology of the Ural Branch of the RAS, Ekaterinburg, Russia
[2]Ural Federal University, Ekaterinburg, Russia

Abstract. Age-related changes in human cardiomyocytes are closely related to cardiac diseases, especially atrial fibrillation. Restricted availability of biological preparations from the human atrial myocardium complicates experimental studies on the aging processes in cardiomyocytes. In this preliminary study, we used available experimental data on the age-related changes in ionic conductances in canine atrial cardiomyocytes to predict possible consequences of similar remodeling in humans using two mathematical models (Courtemanche98 and Maleckar09) of human atrial cardiomyocytes. The study was performed using the model population approach, allowing one to assess variability in the cellular response to different interventions affecting model parameters. Here, this approach was used to evaluate the effects of age-related parameter modulation on action potential biomarkers in the two models. Simulation results show a significant decrease in the action potential duration and membrane potential at 20% of the action potential duration in aging. These model predictions are consistent with experimental data from mammalians. The action potential characteristics are shown to serve as notable biomarkers of age-related electrophysiological remodeling in human atrial cardiomyocytes. A comparison of the two models shows different behavior in the prediction of repolarization abnormalities.

1 Introduction

Atrial fibrillation is one of the most common diseases in the human population and is widespread among seniors. The electrophysiological changes in human atrial cardiomyocytes in the aging heart may be strongly related to arrhythmia occurrence and sustainability. However, experimental observation of age-related changes in human atrial cardiomyocytes is complicated by ethical reasons and the restricted availability of healthy myocardial preparations from the human heart, especially of different ages.

In this preliminary study, we used available experimental data on age-related changes in ionic conductances in canine atrial cardiomyocytes [1, 2] to predict possible consequences

*This work was supported by RFBR #18-015-00368, IIP UrB RAS theme #AAAA-A18-118020590031-8, RF Government Act #211 of March 16, 2013, the Program of the Presidium RAS.
**e-mail: konstantin.ushenin@urfu.ru
***e-mail: soloveva.olga@urfu.ru

Table 1. Parameters of the normal action potential (AP) in human atrial cardiomyocytes. The last column shows parameters that were chosen for the classification of normal and pathological action potential in the current study. APD20, APD50, APD90 – action potential duration at X% of repolarization, APA – action potential amplitude, RMP – resting membrane potential, V20 – voltage at 20% of action potential duration.

article	Sánchez et al. 2014 [6]	Muszkiewicz et al. 2018 [5]	Our populations
APD20 (ms)	1 – 60	0.7 – 11.5	0.7 – 60
APD50 (ms)	6 – 200	4.9 – 34.1	4.9 – 200
APD90 (ms)	190 – 440	68.9 – 142.8	68.9 – 440
APA (mV)	75 – 120	94.6 – 137.5	75 – 137.5
RMP (mV)	-85 – -65	-85.2 – -66.1	-85.2 – -65
V20 (mV)	-35 – 10	–	-35 – 10

of similar remodeling in humans using two mathematical models of human atrial cardiomyocytes: Courtemanche98 [3] and Maleckar09 [4]. The main goal of this analysis was to define the most important electrophysiological biomarkers for age-related changes in human atrial cardiomyocytes. The study was performed using the model population approach [5] to analyze the effects of age-related variation of model parameters on observed model behavior. We also compared the aging model populations based on the two ionic models of atrial cardiomyocytes.

2 Methods

We used Courtemanche98 [3] and Maleckar09 [4] ionic models for analysis of the electrophysiological consequences of cardiomyocyte aging. These models are widely used in computational electrophysiology studies.

There are essential differences between the models. Both models simulate human atrial action potential, but they produce the action potentials of different shapes. This is a result of different definitions of ionic currents in the models.

Both models are based on experimental data on the human atrial myocardium from patients of mostly the senior age group. Particularly, parameters of the ionic currents in the models were derived from experimental data recorded in the myocardium of patients of the following ages: 62±1 years old in [7], 59±2 years old in [8], 61 to 78 years old in [9], 61±1 years old in [10], 63.9±10.7 years old in [11], 41 – 78 years old in [12], and 45 – 71 years old in [13]. Thus, we may consider the two mathematical models as reference models representing human atrial cardiomyocytes of the senior age group.

Almost no systematic data on the action potential in human cardiomyocytes from the adult or young adult group and no data on progressive changes in age-related cardiomyocyte characteristics throughout the aging process are available at the moment. On the contrary, such data is available for the canine atrium, as presented in Table 2. Based on this information, we chose several experimentally defined age-modified parameters of ionic currents to evaluate the effects of their variation on the action potential characteristics in human atrial models. We varied five parameters for the Na^+ current (I_{Na}), transient outward K^+ current (I_{to}), and L-type Ca^{2+} current (I_{CaL}): three maximal conductances g_{Na}, g_{to}, g_{CaL}, and two kinetic parameters of I_{to} - half-maximal voltage $V_{0.5}(I_{to})$, and the slow time constant of the recovery from inactivation $\tau_{slow_i}(I_{to})$.

This study used the population of models methodology to investigate the behavior of the human atrial action potential with accounting for natural variability in cellular activity even within one testing group. An entire population of atrial action potential models was built with

https://doi.org/10.1051/itmconf/20203101004

the variation of each of the five parameters within the range 0–200% from the reference value (100%) assigned in the original articles [3, 4]. Thus, each model from the population can be described by a five-dimensional scaling parameter vector, the values of which are expressed in percentage of the reference values. The population includes 10,000 models, each of which was pre-paced for 99 cycles at 1Hz frequency to reach the steady-state.

The last action potential shape in each model was classified into one of the following groups: normal action potential (AP), pathological AP, early afterdepolarizations (EADs), delayed afterdepolarizations (DADs), as suggested in [14]. The models were experimentally calibrated according to the physiologically plausible ranges of AP biomarkers reported in the literature, as defined in Table 1. Color scales were used to display the phenotype of each tested model sample within the entire population on the planes of paired input parameter space (Figure 1).

To analyze age-related changes in human atrial model behavior, we proposed five aging sub-populations of models selected from the entire model population: Adult I, Adult II, Senior I, Senior II, and Senior III. The groups were sampled as follows. First, we assigned the reference parameters of human atrial models from [3, 4] as the mean values for the Senior I age group, as the reference models were based on the experimental data from patients of around 60 years old. To define mean parameters for younger and older model sub-populations, we varied the five age-related parameters based on the experimental data on the relative changes in the parameters between the Senior I and the Adult I groups and our assumption on the linear change in the parameters in aging. The direction and percentage of age-related modulation in the mean parameter values between the Adult I and Senior I groups was set according to the experimental data from the canine atrial myocardium from 1–3 (Adult 1) and 8 (Senior I) years old dogs (see Table 2). Then we connected the mean points in the 5-dimensional parameter space by the "aging line" used to compute the mean parameters for the intermediate Adult II group between the Adult I and Senior I groups and for the older groups Senior II and Senior III that lay in the opposite direction from the Senior I group on the aging line (see Table 2 for the mean parameter values in the groups expressed in percentage of the reference values in the Senior I, and the lines in the parameter spaces shown in Figure 1).

The age sub-populations of models were sampled from the sphere in the parameter space around the mean parameter vector for the corresponding group at a distance less than 30% of the length of the reference 5-variable parameter vector (see sub-population circles on the 2-parameter planes shown in Fig. reffig:projections). Each sub-population consisted of 1,080 to 2,070 models.

Age-related biomarkers for AP shape distinguishing between the groups were then derived using a naive Bayesian classifier.

3 Results

Figure 1 shows the projections of the five-dimensional space of varied age-related parameters to two-dimensional planes. The color of each point indicates the AP's class according to the used model classification. As can be seen from Figure 1, the two atrial cardiomyocyte models we tested are differently vulnerable for the repolarization abnormalities within the tested parameter range. The much higher percentage of EAD was generated by the Maleckar09 model, and no DAD was produced by the Courtemanche98 model. The distribution of the models between the different classes within the parameter space is also essentially different between the cellular reference models. Particularly for the Maleckar09 model, EADs and DADs are produced in a narrow range of g_{CaL} of high average value, but with a more wider distribution of the parameters of I_{to} compared to the Courtemanche98 model.

Table 2. Experimental data on age-related changes in the transmembrane currents in canine atrial myocardium.

current	parameter	change with age	article	age (year)	atrium
$I_{Ca,L}$	density	↓ 47%	[1]	2-5 & >8	RA
		↓ 43%	[2]	2-2.5 & >8	LA
		↓ 43% ↓ 37%	[15]	1-3 & >8	LA
I_{to}	density	↑ 48%	[1]	2-5 & >8	RA
	$V(0.5)_i$	↓ 15%			
	τ_{fast_i}	↑ 68%			
	τ_{slow_i}	↑ 246%			
I_{Na}	density	↑ 28% ↑ 22%	[16]	2-5 & >8	RA LA

Table 3. The linear model of age-related changes in human atrial cardiomyocytes. The numbers show parameter values in the aging groups expressed in percentage of the reference values assigned for the Senior I age group. Parameter values in other groups are based on the experimental data and assumption on the liner change in the parameters between the aging groups.

		Courtemanche98 Maleckar09	Adult I	Adult II	Senior I	Senior II	Senior III
$I_{Ca,L}$ [1]	$g_{Ca,L}$	0.12375 mS/uF 0.135 mS/uF	189%	144.5%	100%	55.5%	11%
I_{to} [1]	g_{to}	0.1652 mS/uF 0.150 mS/uF	68%	84%	100%	116%	132%
	τ_i		29%	64.5%	100%	135.5%	171%
	$V(0.5)_i$	-43.1 mV -40.5 mV	118%	109%	100%	91%	82%
I_{Na} [16]	g_{Na} P_{Na}	7.8 mS/uF 0.0018 nl/s	78%	89%	100%	111.5%	123%

The series of AP shapes without repolarization abnormalities generated in the age-dependent sub-populations are shown in Figure 2. The linear model of the age-related parameter modulation that we used predicts qualitatively similar changes in the AP characteristics for both Courtemanche98 and Maleckar09 models. The most notable age-related changes are observed for the action potential duration at 90% of repolarization (APD90) and action potential voltage at 20% of APD (V20) (Figure 2), which both decrease with aging. In the Courtemanche98 model, APD90 does not significantly change between Adult I, II and Senior I age groups and linearly decreases between the Senior I and Senior III groups. In the Maleckar09 model, there is a gradual logistic decrease in the APD90. The V20 dependence on age is similar between the models.

There are several differences in the age-related effects on other AP biomarkers between the models. The Maleckar09 model predicts decreasing resting potential during aging, but the Courtemanche98 model does not. Note also, the Courtemanche98 model produces AP with the spike-and-dome morphology in every aging group of models except the Senior III. Conversely, the Maleckar09 models display this AP morphology in only the Senior III age groups of models. Both models show almost no effect of age-related parameter modulation on

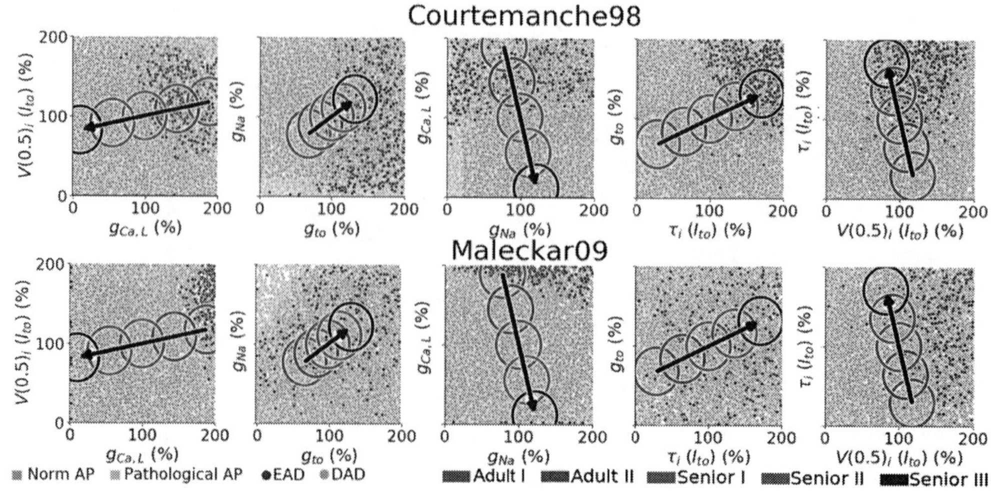

Figure 1. Projection of the five-dimensional parameter space on the two-dimensional planes. Colour points show the classification of APs for the model samples. Age-dependent model sub-populations are shown as the circles on the 2-parameter planes. The black arrow shows the direction of the age-related changes in the models of human atrial cardiomyocytes.

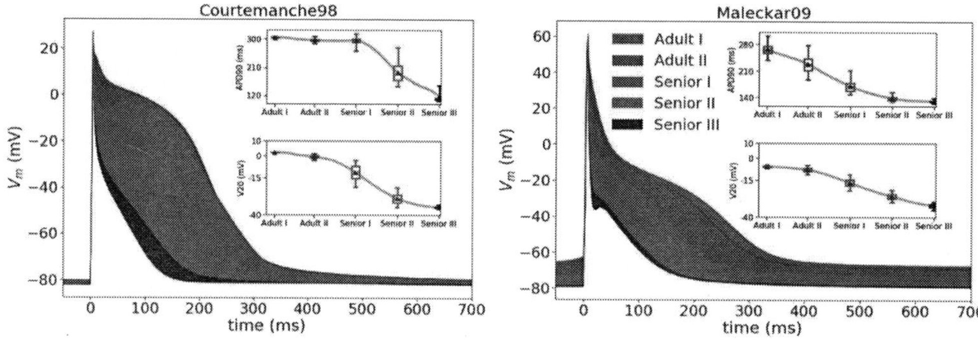

Figure 2. Age-related changes in the action potential of cardiomyocytes. Each color shows one of the age sub-populations. Insert plots show the distribution of APD90 and V20 in the sub-populations.

the absolute value of the AP upstroke and height. Adult II models differ from Senior I models for the V20 parameter: -7.93±1.26 versus -16.58±2.73 mV in the Maleckar09 model, and -0.72±0.99 versus -11.23±4.98 mV in the Courtemanche98 model. Senior I models differ from Senior II models for the APD90 parameter: 170.97±13.97 versus 138.77±5.51 ms in the Maleckar09 model, and 295.12±7.74 versus 195.13±29.94 ms in the Courtemanche98 model. Other parameters of the AP show insignificant differences between the ages in the absolute physical units.

Our goal was to provide most reliable biomarkers of cardiomyocites aging that are based on AP contours. The naive Bayes classifier was able to distinguish model behaviour between the groups and provided the most significant border values related to the age-groups. The values are as follows: Adult II – Senior I, V20=-7.1 (accuracy=99.46%); Senior I – Senior II,

V20=-21.59 (acc.=99.02%), APD90=209.6 (acc.=99.75%); Senior II – Senior III, V20=-31.7 (acc.=85.24%). Other linear rules show accuracy lower than 85%.

4 Discussion

Experimental studies on isolated canine atrial cardiomyocytes show prolongation of the APD90 in aging ([1, 2]). Several studies on the whole human atriums reported an increase in the effective refractory period associated with the age [17, 18]. These observations also suggest an increase in APD90. This data disagrees with our predictions on the APD90 shortening in aging models of human atrial cardiomyocytes suggesting the inter-species difference in the aging processes in atrial cardiomyocytes.

The decrease in V20 in the aging models is consistent with experimental observations in canine atrial cardiomyocytes ([1, 2]). This AP biomarker is shown to closely correlate with parameters of the L-type calcium current and transient outward current, which were directly modified in the aging sub-populations, while APD90 is sensitive to the variation in all transmembrane currents under consideration.

In summary, we suppose that V20 and APD90 are the most trustworthy electrophysiological biomarkers of human atrial cardiomyocyte aging. Both of them show significant changes during in aging in the absolute values of physical units. The change in V20 is consistent with experimental observations. However, APD90 is more suitable for experimental and clinical measurements, because it may be estimated from the activation-recovery interval of the unipolar electrogram.

The presented study shows preliminary results using the model population approach that may be further extended in several directions. Currently, we are working on a more specific definition of age-related model groups. The proposed simple linear approach we used here to simulate age-related changes in model parameters has to be verified in tissue models against available clinical information on age-related changes in intracardiac electrograms and electrocardiograms. Also, we see potential applications of our approach for drug-testing with accounting for age-related changes in cellular activity.

5 Conclusion

In this preliminary study, we used experimental data on the age-related changes in canine atrium cardiomyocytes to build aging populations of models based on the two models of human atrial cardiomyocyte (Courtemanche98 and Maleckar09). The populations of models identify APD90 and V20 as significant biomarkers related to age-dependent changes in human atrial cardiomyocytes. We suppose that distinguishing aging groups in the human population may be classified according to the found biomarkers.

References

[1] W. Dun, T. Yagi, M.R. Rosen, P.A. Boyden, Cardiovascular Research **58**, 526 (2003)

[2] T.Y. Gan, W. Qiao, G.J. Xu, X.H. Zhou, B.P. Tang, J.G. Song, Y.D. Li, J. Zhang, F.P. Li, T. Mao et al., Experimental and therapeutic medicine **6**, 919 (2013)

[3] M. Courtemanche, R.J. Ramirez, S. Nattel, American Journal of Physiology-Heart and Circulatory Physiology **275**, H301 (1998)

[4] M.M. Maleckar, J.L. Greenstein, W.R. Giles, N.A. Trayanova, American Journal of Physiology-Heart and Circulatory Physiology **297**, H1398 (2009)

[5] A. Muszkiewicz, X. Liu, A. Bueno-Orovio, B.A. Lawson, K. Burrage, B. Casadei, B. Rodriguez, American Journal of Physiology-Heart and Circulatory Physiology **314**, H895 (2018)

[6] C. Sánchez, A. Bueno-Orovio, E. Wettwer, S. Loose, J. Simon, U. Ravens, E. Pueyo, B. Rodriguez, PloS one **9**, e105897 (2014)

[7] J. Feng, G. Li, B. Fermini, S. Nattel, American Journal of Physiology-Heart and Circulatory Physiology **270**, H1676 (1996)

[8] A.J. Workman, K.A. Kane, A.C. Rankin, Cardiovascular research **52**, 226 (2001)

[9] P. Gluais, M. Bastide, D. Grandmougin, G. Fayad, M. Adamantidis, European journal of pharmacology **497**, 215 (2004)

[10] Z. Wang, J. Feng, H. Shi, A. Pond, J.M. Nerbonne, S. Nattel, Circulation Research **84**, 551 (1999)

[11] L. Firek, W. Giles, Cardiovascular research **30**, 31 (1995)

[12] Z. Wang, B. Fermini, S. Nattel, Cardiovascular research **28**, 1540 (1994)

[13] A.A. Dawodu, F. Monti, K. Iwashiro, M. Schiariti, R. Chiavarelli, P.E. Puddu, International journal of cardiology **54**, 237 (1996)

[14] A. Muszkiewicz, O.J. Britton, P. Gemmell, E. Passini, C. Sánchez, X. Zhou, A. Carusi, T.A. Quinn, K. Burrage, A. Bueno-Orovio et al., Progress in biophysics and molecular biology **120**, 115 (2016)

[15] G.J. Xu, T.Y. Gan, B.P. Tang, Z.H. Chen, A. Mahemuti, T. Jiang, J.G. Song, X. Guo, Y.D. Li, X.H. Zhou et al., Molecular medicine reports **8**, 1343 (2013)

[16] S. Baba, W. Dun, M. Hirose, P.A. Boyden, American Journal of Physiology-Heart and Circulatory Physiology **291**, H756 (2006)

[17] P.M. Kistler, P. Sanders, S.P. Fynn, I.H. Stevenson, S.J. Spence, J.K. Vohra, P.B. Sparks, J.M. Kalman, Journal of the American College of Cardiology **44**, 109 (2004)

[18] Y.K. Lin, Y.A. Chen, T.I. Lee, Y.C. Chen, S.A. Chen, Y.J. Chen, Circulation Journal **82**, 1237 (2018)

ITM Web of Conferences **31**, 01005 (2020) https://doi.org/10.1051/itmconf/20203101005
Mathematical Modelling in Biomedicine 2019

Wave propagation for a reduced model of blood coagulation

Nicolas Ratto[1,*], *Martine* Marion[1], and *Vitaly* Volpert[1,2,3]

[1]Institute Camille Jordan, UMR 5208 CNRS, University Lyon 1, Villeurbanne 69622, France
[2]INRIA, Université de Lyon, Université Lyon 1, Institute Camille Jordan, Villeurbanne 69200, France
[3]Peoples' Friendship University of Russia (RUDN University), Moscow 117198, Russian Federation

Abstract. Reduced mathematical models of blood coagulation can properly describe Thrombin Generation Test (TGT). Clot growth occurs as a reaction-diffusion wave, and the reduced model studied in this work describes this behaviour, preserving the differences observed between healthy and hemophilia subjects in the TGT.

1 The coagulation process

Blood coagulation is a complex process involving biochemical reactions in plasma initiated at the damaged wall and leading to the formation of fibrin clot that stops bleeding. Thrombin is the central enzyme of the coagulation cascade produced at the initiation and amplification stages of the coagulation reactions. Thrombin generation test (TGT) is one of the most widely used integral laboratory tests of blood coagulation sensitive to vast variety of coagulation disorders [1–4]. It consists in measurement of thrombin concentration in time during *in vitro* coagulation of a plasma sample. In spite of the complexity of blood coagulation kinetics including dozens of substances and reactions between them [5, 6], thrombin generation curves (TGC) are quite simple bell-shaped curves.

Detailed mathematical models of TGT contain dozens of ODEs and many parameters that can hardly be determined for individual patients [7], and the application of such models encounters certain difficulties. In this work, we consider a reduced model of blood coagulation derived from a more complete model [8–10]:

$$\frac{du_{10}}{dt} = (k_1 + k_2T + k_3T^2)(v_{10}^0 - u_{10}) - k_4u_{10}, \tag{1}$$

$$\frac{dT}{dt} = (k_5u_{10} + k_6T + k_7T^2 + k_8T^3)P - k_9T, \tag{2}$$

$$\frac{dP}{dt} = -(k_5u_{10} + k_6T + k_7T^2 + k_8T^3)P, \tag{3}$$

where u_{10} is the activated factor Xa, T is the thrombin concentration, and P the prothrombin concentration. This model can reproduce the TGT for both healthy and hemophilia subjects, and the dependence of thrombin generation curves on the concentrations of blood factors

*e-mail: nicolas.ratto@ec-lyon.fr

https://doi.org/10.1051/itmconf/20203101005

[11]. Let us note that the value of parameter k_1 characterizes the initiation of the reaction at the wall, and k_5 the initial thrombin production. Parameters k_2, k_3, k_6, k_7, and k_8 correspond to the self-accelerating thrombin production due to a positive feedback loop involving activated factor Xa. The simplification of the model based on the approximation of detailed equilibrium for fast reactions allows the reduction of the complete system of equations to the system of three equations with a polynomial dependence on thrombin concentration [11, 12]. Finally, parameters k_4 and k_9 show the action of the anticoagulating factors.

It is known that clot growth occurs as a reaction-diffusion wave [8, 13–15]. The goal of this work is to study this process with model (1)-(3) taking into account the spatial distribution of the involved concentration. In this case, the model becomes as follows:

$$\frac{\partial u}{\partial t} = D\frac{\partial^2 u}{\partial x^2} + F(u). \tag{4}$$

Here $u = (Xa, T, P)$, and F is the vector of reaction rates defined in (1)-(3). The diffusion matrix D is a diagonal matrix with positive diagonal elements.

2 Dynamics of clot growth

2.1 Thrombin generation curves

Thrombin generation curves for healthy subjects and hemophilia patients [11] were fitted by system (1)-(3). Each individual can be characterized by the values of parameters $k_1, ..., k_9$. In the space of three most sensitive parameters k_2, k_6 and k_9, healthy and hemophilia subjects are clearly separated (Figure 1, left).

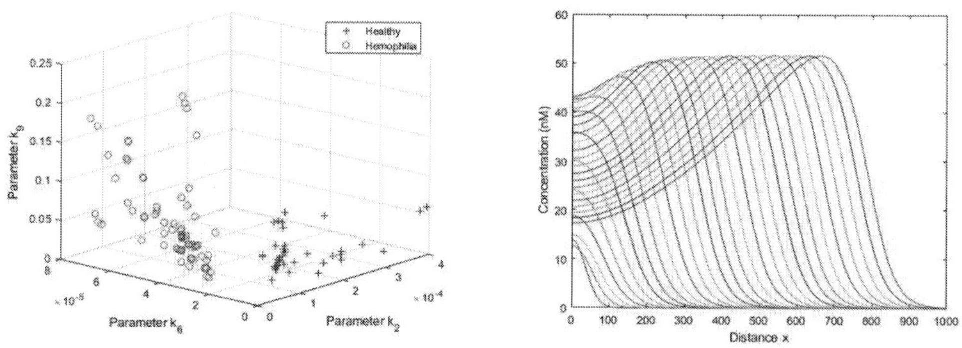

Figure 1. Left: the values of parameters for healthy (crosses) and hemophilia (circles) patients obtained by fitting the experimental thrombin generation curves. Right: wave propagation for a healthy subject ($k_2 = 3.030e^{-4}, k_3 = 1.296e^{-05}, k_4 = 6.785e^{-3}, k_5 = 7.205e^{-06}, k_6 = 1.269e^{-06}, k_7 = 4.749e^{-12}, k_8 = 8.172e^{-12}, k_9 = 1.918e^{-2}$).

We will use below these values of parameters in order to study spatial clot growth in the healthy and hemophilia cases. The nonlinear PDE system is implemented in MatLab with the PDEPE solver. Time integration uses the ODE15s solver with variable time step and variable order [16–18].

2.2 Wave propagation

After the initiation at the vessel wall, clot growth represents a self-sustained wave propagation independent of the initiation stage ($k_1 = 0$). The latter is replaced by the initial condition for the wave propagation:

$$T^0(x) = \begin{cases} T_0 & \text{if } x < x_0 \\ 0 & \text{if } x \geq x_0 \end{cases} . \tag{5}$$

The values of x_0 and T_0 are fixed in all simulations. The initials conditions for the other two functions are taken constant: $u_{10}^0 = 0$ and $P^0(x) = P_0$. An example of wave propagation is shown in Figure 1 (right).

Figure 2 shows the dependence of the wave speed on parameters. The speed increases with the increase of the coefficient k_2 which determines the rate of production of the factor Xa. Let us note that for the given initial condition and for considered values of other parameters, the wave propagation is not observed for $k_2 = 0$. Similarly, the wave speed increases with the value of the coefficient k_5, and it becomes 0 for some positive value. Wave propagation is not observed for lesser values. Since we consider the problem on a positive half-axis, in this case, the wave speed can be negative. Finally, the wave speed decreases with the increase of k_9. This coefficient shows the intensity of thrombin deactivation by antithrombin.

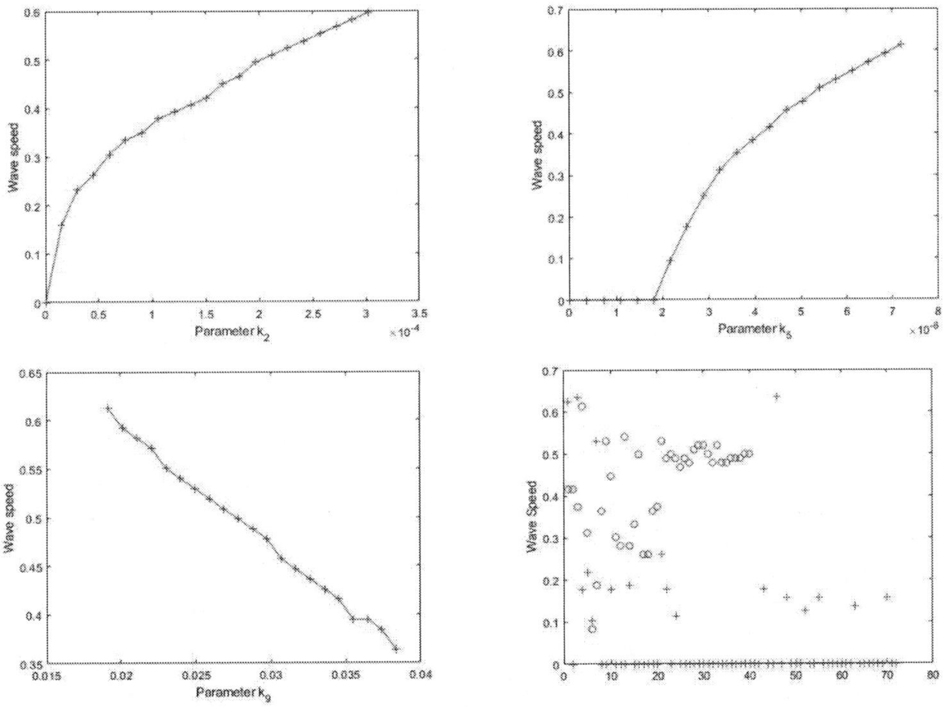

Figure 2. Top figures, and bottom left figure: wave speed depending on parameters k_2, k_5 or k_9, respectively. Bottom right: wave speed for the values of parameters determined by fitting the experimental data. The x-axis shows the number of patient, the y-axis the corresponding wave speed; healthy: circles, hemophilia: crosses.

2.3 Difference between healthy and hemophilia patients

Hemophilia patients are characterized by low concentrations of factor *VIII* or *IX*. In model (1)-(3), the values of parameters k_2, k_3, k_7 and k_8, depending on these concentrations, are also low. Although parameters k_7 and k_8 have a minor impact on the wave speed, parameters k_2 and k_3 are very important. Hence, in the majority of cases, for hemophilia patients we observe wave propagation with a low speed, or no propagation at all. Therefore, clot does not form normally, or it does form not at all.

Figure 2 (bottom right) shows that all healthy subjects have positive wave speeds, which leads to normal coagulation. Most of hemophilia subjects have, instead, a low wave speed or no wave propagation, which leads to bleeding disorders. In few cases, the wave speed for hemophilia subjects is similar to the wave speed for healthy subjects. In fact, it has been observed that some hemophilia patients have normal coagulation.

2.4 The influence of blood flow

Reaction-diffusion waves describe the propagation phase of clot growth but not growth arrest. Two main mechanisms that stop clot growth are determined by the influence of blood flow and by the activated protein C. We discuss here the first mechanism. Taking into account blood flow, we write system (4) in the following form:

$$\frac{\partial u}{\partial t} = D\frac{\partial^2 u}{\partial x^2} + F(u) - \sigma(x)u. \tag{6}$$

The last term in the right-hand side of this equation characterizes the influence of the flow velocity on clot growth washing out blood factors from the clot. The function $\sigma(x)$ show the flow velocity as a function of the distance from the wall. In the first approximation, neglecting the influence of the clot on flow velocity, we can consider a Poiseuille profile. For the same values of parameters as in Figure 1 (right) and $\sigma(x) = ax(L - x)$ we obtain the results shown in Figure 3. The clot growth speed decreases with time, and clot stabilizes at some distance from the wall.

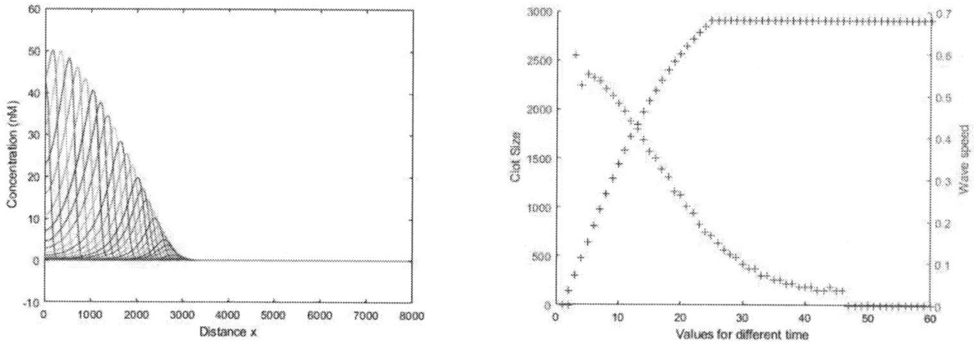

Figure 3. Left: wave propagation for a healthy subjects with a Poiseuille flow $\sigma(x) = ax(L - x)$, where $a = 10^{-4}\mu m^{-2}.s^{-1}$, $L = 8\mu m$. Right: wave speed and clot size as functions of time.

3 Discussion

Existence, stability and the speed of propagation of reaction-diffusion waves of blood coagulation determine the properties of clot growth. We have presented in this work some results obtained by numerical simulations. The underlying mathematical analysis is usually quite involved. For system (4), these questions are not studied. We will discuss here a simplified system where $k_9 = 0$. Assuming that diffusion coefficients in all equations are equal to each other, we can reduce this system to the system of two equations:

$$\frac{\partial u}{\partial t} = d\frac{\partial^2 u_{10}}{\partial x^2} + (k_2 T + k_3 T^2)(v_{10}^0 - u_{10}) - k_4 u_{10}, \tag{7}$$

$$\frac{\partial T}{\partial t} = d\frac{\partial^2 T}{\partial x^2} + (k_5 u_{10} + k_6 T + k_7 T^2 + k_8 T^3)(P_0 - T). \tag{8}$$

This system has two stationary points: $u_{10} = T = 0$ and $T = P_0$, $u_{10} = hv_{10}^0/(k_4 + h)$, where $h = P_0(k_2 + k_3 P_0)$. Hence, it is a monostable case, and, since system (7), (8) is monotone, the waves exist for all values of the speed greater than or equal to the minimal speed [19]. These waves are stable, and the minimal speed admits a minimax representation.

Similar results can be expected for small values of the coefficient k_9. However, if it is sufficiently large, the wave with a positive speed may not exist. Clot growth can be observed for the appropriate choice of initial conditions if the wave exists and its speed is positive. The initial conditions considered above are determined by the values x_0 and T_0. They were taken the same for all patients but the result can depend on the choice of these values. In some healthy cases, the wave exists for all nonzero values of x_0 and T_0, while for some hemophilia subjects, the wave was never observed. In some intermediate cases, increasing x_0 or T_0 supports the convergence to the wave.

Acknkowledgements. V. Volpert was supported by the RUDN University "Program 5-100", and the French-Russian project PRC2307.

References

[1] Hemker, H.C., and S. Béguin, Thrombosis and Haemostasis **74(1)**, 134–38 (1995).

[2] Hemker, H. C. et al., Pathophysiology of Haemostasis and Thrombosis **32(5–6)**, 249–53(2002).

[3] Hemker, H. Coenraad et al., Pathophysiology of Haemostasis and Thrombosis **33(1)**, 4–15 (2003).

[4] Lancé, M.D., Thrombosis Journal **13** (2015).

[5] Mann, K.G., S. Butenas, and K. Brummel., Arteriosclerosis, Thrombosis, and Vascular Biology **23(1)**, 17–25 (2003).

[6] Hoffman, M., Z.H. Meng, H.R. Roberts, and D.M. Monroe, Japanese Journal of Thrombosis and Hemostasis **16(1)**, 70–81, http://joi.jlc.jst.go.jp/JST.JSTAGE/jjsth/16.70?from=CrossRef (2005).

[7] Chelle, P., C. Morin, A. Montmartin, M. Piot, M. Cournil, and B. Tardy-Poncet, Bull Math Biol **80**, 1989–2025 (2018).

[8] Tokarev, A., Yu.V. Krasotkina, M.V. Ovanesov, M.A. Panteleev, M.A. Azhigirova, V.A. Volpert, F.I. Ataullakhanov, A.A. Butilin, Math Model Nat Phenom **1(2)**, 124–137 (2006).

[9] Panteleev, M.A., M.V. Ovanesov, D.A. Kireev, A.M. Shibeko, E.I. Sinauridze, N.M. Ananyeva, A.A. Butylin, E.L. Saenko and F.I. Ataullakhanov, Biophys J **90**, 1489–1500 (2006).

[10] Panteleev, M.A., A. N. Balandina, E. N. Lipets, M.V. Ovanesov, and F.I. Ataullakhanov, Biophys J **98**, 1751–1761 (2010).

[11] Ratto, N., A. Tokarev, P. Chelle, B. Tardy-Poncet, and V. Volpert, Acta Biotheoretica, in press. (2019).

[12] Ataullakhanov, F.I., G.T. Guria, and A.Y. Safroshkina, Biofizika **39(1)**, 97–106 (1994).

[13] Krasotkina, Y. V., E. I. Sinauridze, and F. I. Ataullakhanov, Biochimica et Biophysica Acta (BBA) - General Subjects **1474(3)**, 337–345 (2000).

[14] Pogorelova, E. A., and A. I. Lobanov, Biophysics **59(1)**, 110–118 (2014).

[15] Zarnitsina, V. I., F. I. Ataullakhanov, A. I. Lobanov, and O. L. Morozova, Chaos: An Interdisciplinary Journal of Nonlinear Science **11(1)**, 57 (2001).

[16] Shampine, L. F. and M. W. Reichelt, SIAM Journal on Scientific Computing **18**, 1–22 (1997).

[17] Shampine, L. F., M. W. Reichelt, and J.A. Kierzenka, SIAM Review **41**, 538–552 (1999).

[18] Skeel, R. D. and M. Berzins, SIAM Journal on Scientific and Statistical Computing **11**, 1–32 (1990).

[19] Volpert, V., *Elliptic Partial Differential Equations: Volume 2: Reaction-Diffusion Equations* (Birkhauser, Basel, 2014).

The multiscale simulation of apical myocardial infarction and shape variation of the left ventricle of the heart

Fyodor Syomin[1,2,*], *Albina* Khabibullina[3], *Anna* Osepyan[3], and *Andrey* Tsaturyan[1]

[1]Institute of Mechanics, Lomonosov Moscow State University, 1 Mitchurinski pr-t, Moscow, 119192, Russian Federation
[2]Peoples' Friendship University of Russia (RUDN University), 6 Miklukho-Maklaya St, Moscow, 117198, Russian Federation
[3]Faculty of Mechanics and Mathematics, Lomonosov Moscow State University, 1 Leninskie Gory, Moscow, 119991, Russian Federation

Abstract. A multiscale model of the cardiovascular system (CVS) in which the left ventricle (LV) of the heart was approximated by an axisymmetrical thick-wall body made of transversely isotropic incompressible material was used to simulate the performance of the heart with apical myocardial infarction (MI). The material model reproduced mechanical properties and calcium regulation of active tension in cardiac muscle. The changes in the LV strain and the reduction of the LV stroke volume and arterial blood pressure obtained in the MI simulations were similar to those observed in patients with the apical MI. In contrast to the decrease in heart performance in the MI simulations, the simulation of changes in the LV shape from "normal" to a spherical or conical one revealed only slight changes in haemodynamics provided that the LV preload and the mass of the LV wall were kept constant.

1 Introduction

Mathematical modelling is a useful tool for understanding how the heart works in health and disease [1]. Multiscale models could be used in clinics for more detailed diagnostics and the choice of adequate treatment of heart diseases. In particular, myocardial infarction (MI) is a heart disease of interest for computer simulation [2]. Earlier we had developed a model of the cardiovascular system with a focus on the left ventricle (LV) of the heart [3]. The LV was represented by a multiscale model describing the macroscopic pumping function of the ventricle considering the dependence of stress-strain relationships at the tissue level on the microscopic processes at the cell level [4]. The advantage of the model compared to those with a simplistic description of myocardium mechanics is a detailed description of the interactions of contractile and regulatory proteins in cardiac cells [5], that is important for accurate simulation of the heart mechanics.

Our multiscale model of the LV within CVS described by a lumped parameter model reproduces the time-courses of haemodynamic variables (pressures in the LV, aorta, veins, volume of the LV) for an averaged "healthy" patient, and the effects of some arrhythmias and heart valve diseases [3]. Here we report the result of simulation of the effects of apical

*e-mail: f.syomin@imec.msu.ru

MI on the LV performance and geometry during a heartbeat. Some heart diseases, including ischemic and non-ischemic LV dilatation and MI, are often accompanied by changes in the LV shape. Surgical ventricular restoration (SRV) is often used to restore the elliptical shape of LV in patients with MI [6] resulting in an increase in the ventricular performance reportedly. On the contrary, the decrease in heart performance was reported for an early postoperative period after SVR of apical MI resulting in more spherical shapes of LVs [7, 8]. The question whether the LV shape changes *per se* affect the LV performance and should be eliminated surgically is under discussion [6, 7, 9]. To address this question, we simulated the effects of changes in the LV shape at the same mass and mechanical properties of the LV wall on the LV function. The LV shape was characterized by the sphericity and conicity indices used in clinical practice [10].

2 Methods

2.1 Mathematical model of the LV within CVS

The key element of our CVS model is a multiscale model of the LV. The ventricle was approximated by a body of rotation close to a semi-ellipsoid [11]. Every point of the ventricular wall in the rectangular coordinate plane was associated with curvilinear coordinates γ (the point position across the wall equal zero at subendocardium and one at subepicardium) and ψ (the point angular position equal zero at the base of the LV and $\pi/2$ at its apex). The corresponding equations were

$$x = (r_{in} + \gamma \cdot (r_{out} - r_{in})) \cdot (\varepsilon \cos\psi + (1-\varepsilon)(1-\sin\psi)),$$
$$y = \left(h_{in} + \gamma \cdot h_{apex}\right) \cdot (1-\sin\psi) + (1-\gamma) \cdot h_{apex}. \tag{1}$$

Here, parameters r_{in}, r_{out}, h_{in}, h_{apex}, and ε set a preliminary end-systolic geometry, in particular, inner and outer radii at the base, the axial size of the ventricular cavity, the thickness of the apex, and a variation in the curvature of the wall (the shape is described by a cone at ε equals zero).

The myocardium of the LV wall was assumed to be incompressible transversely isotropic material in that non-isotropic stress of axial forces applied along the fibres of cardiac muscle was described by our kinetic model [5]. The kinetic model was set by a system of ordinary differential equations (ODEs) for variables determining several states of contractile and regulatory proteins. In all numerical experiments presented in the paper, the myocardium fibres were oriented in a spiral manner with helical angle changing from 80° at the inner LV wall to −55° at the outer wall.

The other three chambers of the heart, as well as systemic and pulmonary circulations, were described by a new lumped parameter model, which was also set by ODEs for pressures P and volumes V of several segments of the vascular bed, considered as elastic reservoirs, and for the flows Q between them. Inertial and hydraulic flow resistances (R and L) and compliances C of each reservoir were assumed to be constant parameters. The general forms of the equations were

$$C\frac{dP}{dt} = Q_{in1} - Q_{out1}, \frac{dV}{dt} = Q_{in2} - Q_{out2}, L\frac{dQ}{dt} + R \cdot Q = P_1 - P_2. \tag{2}$$

Atria and the right ventricle were treated as viscoelastic bodies pressure-volume relationships for which included an activation function depended on time and the chamber volume. The equations for the pressures caused by an active stress in the walls of the chambers had the following form

$$\tau\frac{dP}{dt} + P = F_{Act}\cdot\left(\mu\frac{dV}{dt} + E_{Act}\right),$$
$$F_{Act} = e_{Act}(t)\cdot\left(k_V + (1-k_V)\cdot\left(\frac{V-V_{min}}{V_{max}-V_{min}}\right)\right). \quad (3)$$

The first expression in 3 is an analogue of Force-Velocity relationship for the muscle with non-dimensional force F_{Act} and maximal isovolumic steady-state pressure E_{Act}. The second expression states how F_{Act} depended on activation function e_{Act} and, in order to reproduce Frank's law of the heart, on the chamber volume. The lumped parameter model was described in our recent paper [3].

The equilibrium equations for the LV were related to the lumped parameter model by the volume of the LV cavity V_{LV} and LV pressure P_{LV}, which was included in boundary conditions for the inner wall of the ventricle. Some reasonable initial conditions for the lumped parameter model and the kinetic model of muscle contraction were chosen. The LV contractions were initiated by the periodic simultaneous influx of calcium ions to the cells of the ventricular wall and had been performed until the steady periodical solution had been obtained. The details of the model and its numerical implementation are given in our previous works [3, 4].

2.2 Algorithm of calculation of the shape indices

The shape indices were calculated by an algorithm suggested in [10]. Both sphericity and conicity indices S_i were set by the expression $S_i = 1 - A/A_{max}$, where A is an area between the curves of shape description signals ρ_{LV} and ρ_{Ref} which were plotted on a normalized angular coordinate (zero at the ventricular apex, one at the base). In our case of axisymmetric geometry, the functions ρ_{LV} and ρ_{Ref} had simplified forms. The first function is the distance from the ventricular axis to the point at the endocardial surface normalized by its maximal value and multiplied by the halved aspect ratio of the LV equatorial cross section; the second one is the signal of the reference shape that is the hemisphere with the radius of 0.5 and the centre at (0.5, 0), for the sphericity index, or it is the straight line $y = a \cdot x$ for the conicity index ($a = r/h$ is a halved aspect ratio of the ventricle). A_{max} is an area of the whole coordinate plane equal to 0.5 by default. The end-diastolic coordinates of the inner nodes of our simulated LV were used for ρ_{LV} calculations. Due to the axial symmetry we could use normalized longitudinal coordinates of the nodes as the parameter of ρ_{LV} instead of angular coordinates.

3 Results and discussion

3.1 Apical myocardial infarction

Our model allows us to simulate local defects in properties of the LV myocardium. The most straightforward dysfunction that can be simulated is a regional axisymmetric myocardium infarction (MI). The simulated apical MI was set by a local decrease in the additional factor F_{inf} (multiplier of the active tension) in the apical region as follows

$$F_{inf} = \begin{cases} 1, \psi < \psi_1 \\ k_{inf} + \left(1 - k_{inf}\right)\cdot\left(\frac{\psi_2-\psi}{\psi_2-\psi_1}\right)^2, \psi_1 \le \psi \le \psi_2 \\ k_{inf}, \psi > \psi_2 \end{cases} \quad (4)$$

The passive mechanical properties of the LV myocardium were remained unchanged. Simulation of the apical infarction was performed with the values of the parameters k_{inf}, ψ_1, ψ_2

being equal zero, $\pi/4$ and $3\pi/10$, respectively. Relative volume of the infarcted tissue was near 14% of the total volume of the ventricular wall.

The end-systolic and end-diastolic shapes of the normal and MI LVs with colour maps for sarcomere length are shown in figure 1. The results for the MI LV demonstrate significantly

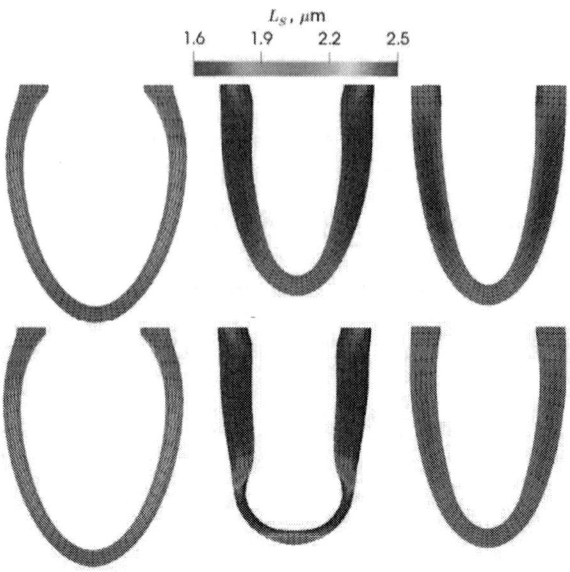

Figure 1. The shape and sarcomere length for the models of "normal" LV (top) and the LV with apical MI (bottom). End-diastolic (left), end-systolic (middle) LVs and the shape at the beginning of diastole (right) are shown

non-uniform fibre strains in the apical region at the end of systole including, in particular, abnormally large strains of endocardial fibres at the apex. One of the clinical parameters for estimating ventricular dysfunction is so called global longitudinal strain (GLS). In medical practice GLS is measured by tracking the deformations of several segments of the ventricular wall in an ultrasound image [12]. It was found out that GLS is usually decreased in patients with heart failure, and the GLS measurements are more reliable than measurements of ejection fraction [13] for the prediction of the heart failure. A correlation of GLS with MI size was reported [14], and the mean MI size of 15% corresponded to the mean GLS of −13% while normal GLS was assumed to be equal −20%. Again MI size had better correlation with GLS than with ejection fraction. The results of our simulation of the MI LV also demonstrate a notable decrease in magnitude of GLS (−9%) compared to the "healthy" LV (−17%).

The resulting time-courses of the main haemodynamic variables, presented in figure 2, show impaired ventricular performance: decreases in stroke volume (from 81 to 72 mL), ejection fraction (from 64% to 56%) and maximal pressures in the LV (from 130 to 117 mm Hg) and aorta (from 122 to 109 mm Hg). However, the decrease in heart performance was not very significant considering quite wide region of MI. That might be caused by limitations of our model. It seems that the presence of the right ventricle and its connection to the LV in the apical region could be crucial for stress and strain distribution in the region. Moreover, our current model does not consider possible remodelling of the LV after MI commonly observed in medical practice. In particular, it was estimated both experimentally and numerically that the MI region becomes stiffer and the distribution of the cardiac fibres orientation changes

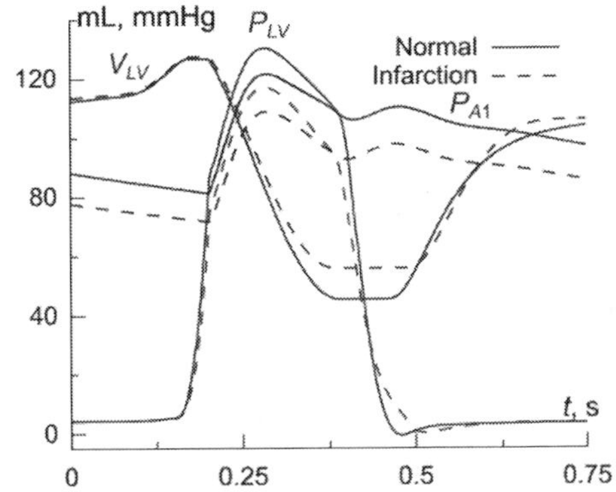

Figure 2. Simulated haemodynamics of the LV with apical MI. The time-courses of the LV volume, pressure, and aortic pressure for the "normal" LV (solid lines) and for the MI LV (dash lines) are presented

Table 1. Shape indices of the simulated LVs

	"Normal"	Spherical	Conical
Sphericity index	0.73	**0.83**	0.58
Conicity index	0.76	0.58	**0.84**

significantly under the remodelling [15]. These changes in the myocardial stiffness and fibre orientation were not taken into account in our model.

3.2 Effects of LV shape

We have performed simulations to examine numerically the dependence of the LV performance on its shape. Geometrical parameters r_{in}, r_{out}, h_{in}, and h_{apex} were set up in such way that the cavity volume and the wall thickness of the LV preliminary end-systolic geometry would be the same as they were for the "normal" LV, while the relation of the radius of the ventricular cavity to its length would change drastically. Then we scaled the whole geometry to obtain the "normal" values of the volume of the ventricular wall. We also slightly changed initial preload (initial pressure values in pulmonary and systemic veins) in order to obtain 'healthy' end-diastolic LV pressures observer in simulations of the "normal" LV. The end-systolic and end-diastolic geometries of the spherical and conical LVs with a colour legend for sarcomere lengths are shown in figure 3. These results demonstrate less uniform distribution of fibre strain across the ventricular wall for the LVs with high conicity or sphericity indices than for the "normal" one. The sarcomeres shortened more significantly during systole of the spherical LV, while the subendocardial fibres underwent larger deformations during diastole. For conical LV one can notice less uniform end-systolic distribution of sarcomere lengths and very small strains in the apex. That means, considering length-dependent activation of the muscle, that muscle fibers in the middle-part equatorial region of the conical LV sustained high non-uniform stress, while the apex did not contract properly.

The signal curves for each ventricle geometry shown in figure 3 are presented in figure 4, while calculated sphericity and conicity indices are presented in table 1.

The time-courses of haemodynamic variables for these three ventricles are shown in figure 5. In spite of the changes in geometry described above, the stroke volumes of the LV at the "normal" level of preload (end-diastolic LV pressure of 5.3 mm Hg) were approximately

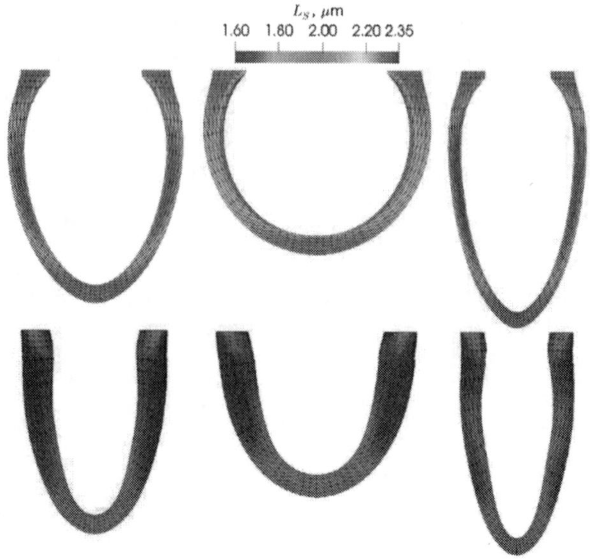

Figure 3. The end-diastolic (top) and end-systolic (bottom) geometry and sarcomere length for the "normal" (left), spherical (middle) and conical (right) LVs

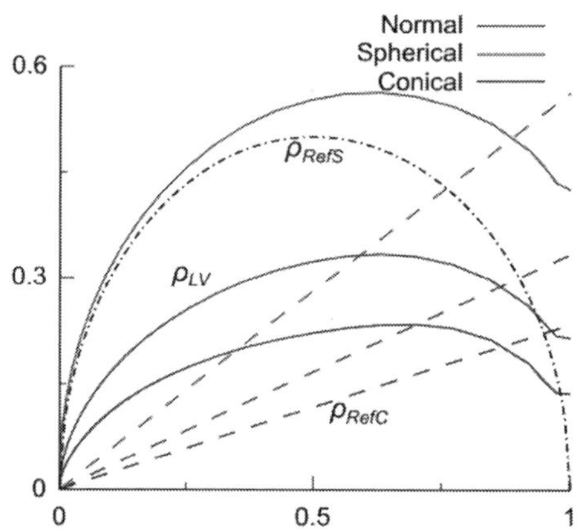

Figure 4. Shape signal curves. Solid lines are the signals for the LVs with "normal", spherical, and conical shapes. Dotted lines are the reference conical shape signals, and the dash-dotted line is the reference spherical signal

preserved (81 mL for the "normal" LV, 78 mL for the spherical LV, and 85 mL for the conical LV), as well as ejection fractions (64% for the "normal" LV, 60% for the spherical LV, 63% for the conical LV). The systolic aortic pressures were slightly increased for the conical LV (126 mm Hg), and decreased for the spherical one (117 mm Hg) compared to the "normal" LV (122 mm Hg). We believe that the decreases in stroke volume observed in spherical LVs after SVR reported in [7, 8] could be caused by the decrease in the mass of the active myocardium in ventricular wall after surgery. Also we should notice that again, as in simulations of MI, we did not consider redistribution of the orientation of muscle fibres through the ventricular

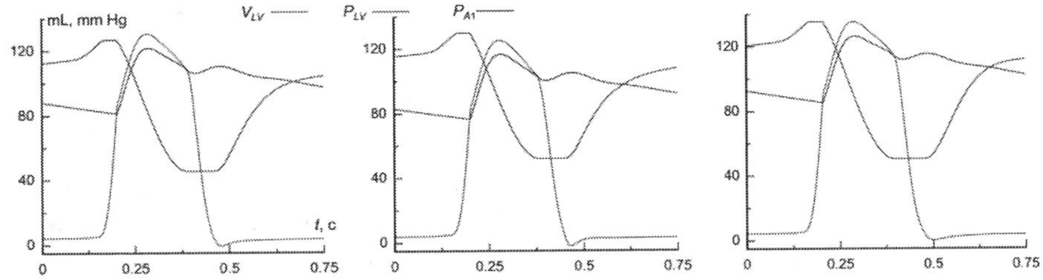

Figure 5. Simulated haemodynamics of the LV with different shape. The time-courses of the LV volume, pressure, and aortic pressure for the "normal" LV (left), spherical LV (middle), and conical LV (right) are presented

wall that was reported to have significant effects on the performance of dilated spherical LVs [6].

4 Conclusions

Our simulation of apical LV MI agrees with clinical data showing a reduction in the stroke volume and ejection fraction after MI involving 15% of LV surface in the apical region [14]. A substantial difference between the end-systolic and early diastolic LV shape is another feature of the simulated MI. Importantly, the simulation predicts a significant change in longitudinal strain of the LV apex and in the shape of the LV cavity during isovolumic relaxation (figure 1). Movement of the LV wall during isovolumic relaxation was suggested to be an independent predictor of heart failure [16].

The results of our simulations showed that changes in the LV shape at the constant mass of its wall did not cause any significant reduction in stroke volume and ejection fraction if the distribution of fibre orientation within the wall remained similar to that in the "healthy" LV. On the other hand, the distribution changes upon LV dilatation so that the helical angle of the subendocardial and subepicardial fibres decreases [6]. As this phenomenon was not considered in our simulations, further work is needed to simulate the effect of the distribution on the LV function.

The reported study was funded by RFBR according to the research project # 18-31-00065.

References

[1] N. A. Trayanova, J. Constantino, V. Gurev, Am J Physiol Heart Circ Physiol **301(2)**, H279–86 (2011)
[2] K. Mangion, H. Gao, D. Husmeier, et al., Heart **104(7)**, 550–557 (2018)
[3] F. A. Syomin, M. V. Zberia, and A. K. Tsaturyan, International Journal for Numerical Methods in Biomedical Engineering **e3216** (2019)
[4] F. A. Syomin and A. K. Tsaturyan, Russian Journal of Numerical Analysis and Mathematical Modelling **32(5)**, 327–337 (2017)
[5] F. A. Syomin and A. K. Tsaturyan, Journal of Theoretical Biology **420**, 105–116 (2017)
[6] G. Buckberg, C. Athanasuleas, J. Conte, Nat Rev Cardiol **9**, 703–716 (2012)
[7] Y. Wang, C. Q. Gao, G. Wang, Y. S. Shen, Chin Med J **130**, 1429–1434 (2017)
[8] L. Ch. Lee, J. F. Wenk, L. Zhong, et al., J Appl Physiol **115**, 136–144 (2013)
[9] R. H. Jones, E. J. Velazquez, R. E. Michle, et al., N Engl J Med **360**, 1705–1717 (2009)

[10] F. Maffessanti, R. M. Lang, C. Corsi,x V. Mor-Avi, and E. G. Caiani, Ultrasound in Med. & Biol. **35(12)**, 1953–1962 (2009)

[11] S. F. Pravdin, V. I. Berdyshev, A. V. Panfilov, et al., Biomed Eng Online **12**, 54 (2013)

[12] D. Medvedofsky, K. Kebed,L. Laffin, et al, Echocardiography **34**, 365–370 (2017)

[13] S. Karlsen, T. Dahlslett, B. Grenne, et al., Cardiovasc Ultrasound **17**, 18 (2019)

[14] G. Joseph, T. Zaremba, M. B. Johansen, et al., Echo Res Pract·**6(4)**, 81–89 (2019)

[15] D. Mojsejenko, J. R. McGarvey, S. M. Dorsey, et al. Biomech Model Mechanobiol **14**, 633 (2015)

[16] P. Brainin, S. Haahr-Pedersen, M. Sengeløv, et al., Int J Cardiovasc Imaging **34**, 751–760 (2018)

Algorithm of optimization of fractionated radiotherapy within its combination with antiangiogenic therapy by means of mathematical modeling

Maxim Kuznetsov[1,2,*] and *Andrey* Kolobov[1,2]

[1]Division of Theoretical Physics, P.N. Lebedev Physical Institute of the Russian Academy of Sciences, 53 Leninskii Prospekt, Moscow 119991, Russian Federation
[2]Peoples' Friendship University of Russia (RUDN University), 6 Miklukho-Maklaya St, Moscow, 117198, Russian Federation

Abstract. Antiangiogenic therapy is the anti-tumor treatment, that leads to the cessation of blood vessels formation and subsequent nutrient deprivation of the tumor. One of its frequently occurring effects is the transient improvement of tumor oxygenation, which lasts several days and leads to increased efficiency of successive exposure to radiation. It is followed by an escalation of hypoxia, and therefore the addition of antiangiogenic therapy to the whole course of fractionated radiotherapy has an ambiguous effect. Previously we have developed a mathematical model of such combined anti-tumor treatment. Herein we present an algorithm of optimization of radiotherapy fractionation, which aim is to find the most optimal distribution of irradiation doses in order to increase the efficiency of such combined treatment. We demonstrate an example of its work and discuss its further application.

1 Introduction

Radiotherapy (RT) is one of the classical options of cancer treatment. It acts by causing damage to cell DNA, and the presence of oxygen sufficiently enhances this process. In vitro experiments show, that radiosensitivity of cells in air and under hypoxia may vary up to three-fold [1]. This fact has led to the active search of methods for improvement of tumor oxygenation during RT. One of them is the combination of RT with antiangiogenic therapy (AAT), which is aimed at ceasing angiogenesis, i.e., the formation of new blood vessels, which supply tumor with nutrients. It was shown, that AAT can cause transient alleviation of intratumoral hypoxia, which lasts for several days at the beginning of the therapy [2]. This phenomenon was explained by the fact, that AAT causes normalization of structure of tumor microvessels and subsequent increase of their perfusion [3].

However, due to transient nature of this phenomenon, its manifestation does not guarantee that addition of AAT would enhance the overall efficiency of the whole course of fractionated RT. Moreover, the ultimate result – and primary goal – of AAT is the nutrient deprivation of a tumor, which means escalation of hypoxia in the long run [4]. Therefore, it is not surprising that preclinical and clinical studies demonstrate ambiguous results concerning the

*e-mail: kuznetsovmb@mail.ru

efficiency of such combined treatment [5, 6]. In our recent work [7], we presented a physiologically justified mathematical model of tumor growth and treatment with RT and AAT and we demonstrated, that tumor radiosensitivity should be one of the factor which define the outcome of addition of AAT to RT: under its low values inclusion of AAT may prolong patient's survival; however, under high tumor radiosensitivity, AAT may compromise curative effect of RT. In that work we simulated the clinical scheme of RT fractionation, in which all doses of irradiation are equal. However, due to the influence of AAT on oxygen dynamics, it is possible, that increased efficiency of such combined treatment can be achieved by altering the distribution of irradiation doses over time. In order to investigate this possibility, we have developed an algorithm of optimization of radiotherapy fractionation with a limited total irradiation dose, which in practice is dictated by the negative effect of RT on normal tissues. Herein we present this algorithm, we demonstrate the result of its work within a model simulation under low tumor radiosensitivity and we discuss its further application.

There exist a number of works on optimization of antitumor tretament by means of mathematical modeling. However, mathematical modeling of combined radiotherapy and antiangiogenic therapy is nowadays by itself a little-studied field. The existing works on this topic consider only purely phenomenological dependencies, which significantly reduces their potential practical value [8, 9]. To the best of our knowledge, there is only one work, in which an attempt to optimize such treatment was made [10]. It as well considers a phenomenological model wherein antiangiogenic therapy directly affects the parameters of tumor growth and radiosensitivity of tumor cells, and oxygen is not accounted for. The existing works on optimization of monoradiotherapy by the means of mathematical modeling mainly consider spatial redistribution of a single course of radiotherapy, that is dictated by clinical needs [9, 11, 12]. There exist several works on optimization of time scheduling of chemotherapy, which as well consider non-spatially distributed phenomenological models and rely on optimal control techniques [13, 14]. Such approach would face tremendous difficulties when dealing with complex spatially-distributed physiologically-based models. In contrast, the approach presented herein can be integrated into the models of any reasonable complexity.

2 The model

2.1 Model of tumor growth and treatment

The model of tumor growth that we use is a continuous spatially-distributed model of reaction-diffusion-convection type, governed by partial differential equations. For full description of the model, including its equations and the values of parameters, we refer the reader to our work [7]. Herein we provide only the brief summary of the model.

The scheme of interactions between the model variables is presented in Fig. 1. The tumor grows in normal tissue, which consists of normal cells and includes preexisting normal microvasculature. Glucose and oxygen flow into tissue from microvasculature and are consumed by tumor and normal cells. Proliferating tumor cells push out surrounding tissues, which is one of the drivers of tumor growth. Microvasculature degrades within the tumor due to elevated solid pressure as well as chemical factors, which are not considered explicitly [15]. Under significant fall of glucose level, tumor cells pass into quiescent state, in which they, as well as normal cells, may die under long depletion of oxygen, turning into necrosis. Quiescent cells migrate throughout the tissue, which represents another driver or tumor growth. As these cells experience metabolic stress, they secrete vascular endothelial growth factor, or VEGF, which stimulates angiogenesis, i.e., formation of angiogenic capillaries [16]. They are described by a separate variable, which is introduced separately to account for "abnormality" of structure of tumor angiogenic capillaries, that results in their

https://doi.org/10.1051/itmconf/20203102001

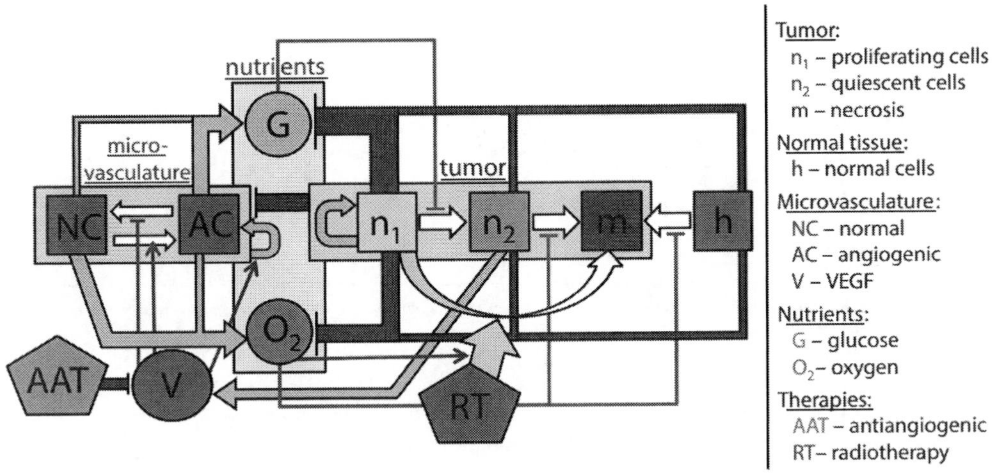

Figure 1. Scheme of the model of tumor growth in tissue with account of radiotherapy and antiangiogenic therapy. White arrows indicate transitions between variables, green arrows denote stimulating effects that increase the value of a corresponding variable or the intensity of a corresponding transition, red lines designate inhibitory interactions.

increased permeability for glucose as well as decreased blood flow and subsequent decreased oxygen inflow in tissue through a single capillary. Also, in presence of VEGF the structure of normal capillaries becomes abnormal and restores in its absence [17]. Antiangiogenic therapy (AAT) is modelled by instantaneous elimination of VEGF from the tissue till the end of a model simulation. Radiotherapy (RT) leads to direct instantaneous death of part of malignant cells – sensitivy of proliferating cells to it being an order of magnitude higher than that of quiescent cells. For RT description we rely on the classical linear-quadratic model, which is widely used in clinical applications [18]. The dependence of tumor cell radiosenstivity on oxygen concentration has been deduced from experimental data in [19].

The growth of tumor is simulated in one-dimensional region, with left boundary representing the center of the tumor, and right boundary being free and thus reflecting the assumption that the normal tissue does not hinder the tumor growth. Figure 2(a) shows the distributions of the model variables corresponding to the growth of an untreated tumor. Functional capillaries inside the tumor are few, and most of angiogenic capillaries are located in the peritumoral area, where they lead to enhanced supply of nutrients to the tumor rim. Therefore, proliferating cells are concentrated there, while the main tumor mass consists of quiescent cells and necrosis.

Figure 2(b) shows the distributions of the model variables on the first day of RT, when some of the tumor cells have already turned into necrosis. Figure 2(c) refers to the same moment under combined RT and AAT, when the latter has already led to normalization of the absolute majority of capillaries, that has resulted in increase of oxygen inflow inside the tumor. In its turn, this has increased the efficiency of the first irradiation, which can be seen from the decrease in density of proliferating tumor cells.

2.2 Algorithm of optimization of radiotherapy fractionation

The task of finding the optimal protocol for fractionated RT under its combination with AAT was formalized as follows. The initial conditions are the distributions of the model variables

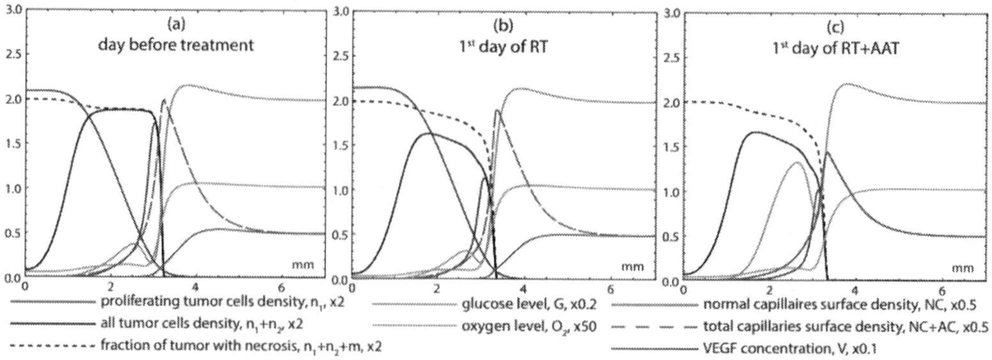

Figure 2. Profiles of the model variables: a) on the day before the treatment; b) on the first day of monoradiotherapy; c) on the first day of combined radiotherapy and angiogenic therapy.

at the time of the initiation of therapy, which takes place when the tumor radius has just reached 3 mm. Both RT and AAT start simultaneously, which is a standard clinical practice. The aim of the search is to find the distribution of fractional doses of irradiation, which would provide the maximum time of the tumor growth till the endpoint – radius of 7 mm – under the conditions, dictated by the clinical practice: 1) the total irradiation dosage is 60 Gy, 2) no more than one irradiation is carried out once every 24 hours, 3) the maximum number of fractions is 30 (or, equivalent to points 2 and 3, there are exactly 30 daily fractions, some of which may have zero dose).

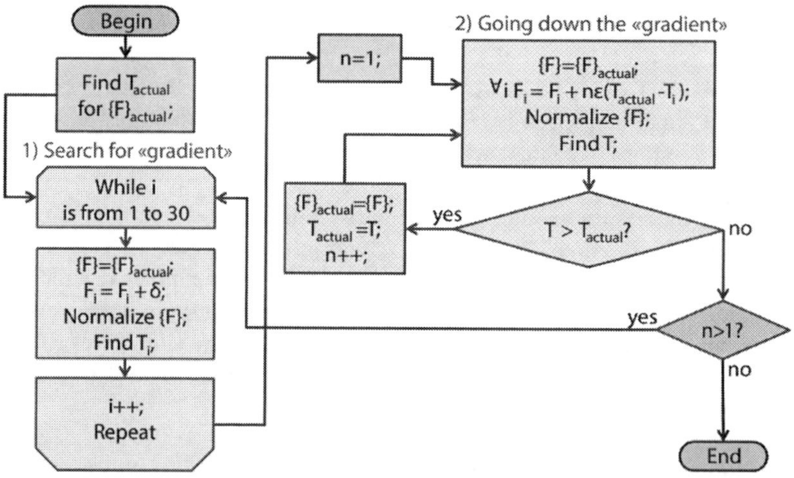

Figure 3. Block-scheme of the algorithm of optimization of radiotherapy fractionation within the utilized model of tumor growth and treatment. See explanations in the text.

For solution of this problem, the basic algorithm, summarized in Fig. 3, was developed and implemented within a single C++ code along with the code for the model of tumor growth and treatment. The utilized algorithm resembles the classical gradient descent method. For the initial scheme of RT, $F_{actual} = \{F_1, F_2, ..., F_{30}\}$, a scheme with 30 daily fractions with doses of 2 Gy is taken, which follows a standard clinical scheme with the exception that we

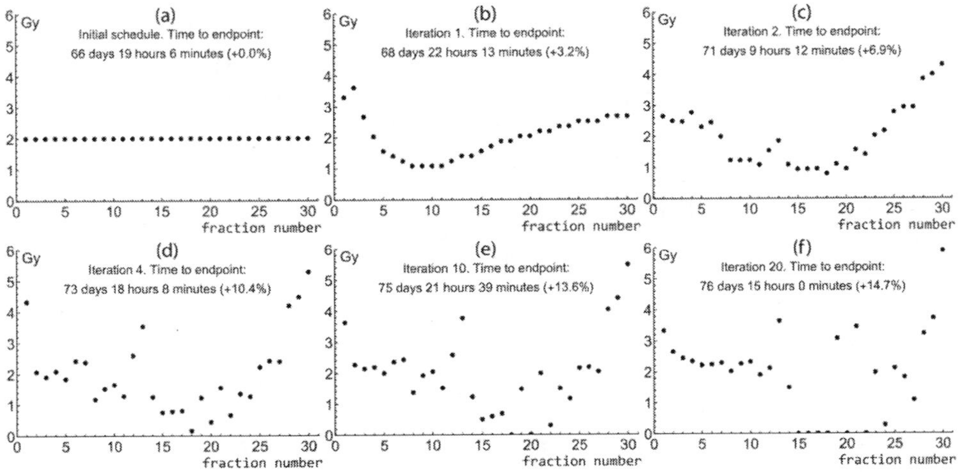

Figure 4. (a) The initial schedule of radiotherapy fractionation, (b-e) intermediate results and (d) final result of the work of the optimization algorithm. The designated time is measured from the moment, when tumor radius is 3 mm and the therapy begins, till the moment, when tumor radius reaches 7 mm.

neglect the fact that irradiations usually do not take place on days-off. The time, which it takes for the tumor to achieve the endpoint from the beginning of the therapy, T_{actual}, is measured. Further the following steps are iteratively repeated. The first step is the search for "gradient", during which tumor growth times T_i are measured for each of the 30 schemes of RT, which are obtained by adding the value of δ, which is the parameter of the algorithm, to one of the fractional irradiation doses, F_i, $i = 1, 2, ..., 30$, and further normalizing the fractional doses so that their sum is 60 Gy. The second step is the descent along the "gradient", wherein a new scheme is introduced, in which each of the fractional doses is obtained by adding a value, proportional to $T_{actual} - T_i$ to F_i, $i = 1, 2, ..., 30$ (initial coefficient of proportionality ϵ is another parameter of the algorithm), and then all the fractional doses are once again normalized so that their sum is 60 Gy. Then the coefficient of proportionality is increased as long as the tumor growth time from the start of the treatment to the endpoint, T, increases. The scheme, which provides the longest T in this step, replaces F_{actual} and these two steps repeat while the first scheme, obtained during the second step, does lead to additional delay in tumor growth time.

3 Results and discussion

The results of the work of the algorithm under a basic set of parameters of tumor model (see [7]) and parameters of optimization model $\delta = \epsilon = 0.05$ is illustrated in Fig. 4. The first figure represents the initial RT scheme. The others refer to the schemes, obtained at the end of the first, second, fourth, tenth and twentieth iterations, the latter scheme being the final result.

During the simulation, the utilized algorithm proposes the following modifications of the RT scheme. At the first iteration it suggests to enhance the first three doses at the expense of the middle ones and thus to take advantage of the transient hypoxia alleviation window. The several final doses also increase, and the next iteration suggests their further sharp growth, while the alterations in the first doses are on the contrary smoothed out. These two iterations provide approximately 7% increase in the tumor growth time T. During the consequent two iterations, the very first fractional dose isolates by achieving the value of 4 Gy, while the values of its neighbors fall down close to their initial value of 2 Gy. The twelfth, thirteenth and three of the last doses also rise up above their neighbors, and all these modifications add up about 3.5% of increase in T. Each of the following sixteen iterations provides only slight alteration in the RT schedule, however, in total they add up more than 4% to T, leading to overall almost 15% gain in tumor growth time. In the RT schedule, eventually suggested by the algorithm, the first several doses line up in a smooth curve, which begins with the first irradiation dosage of about 3 Gy, with the following doses slightly falling close to the initial value of 2 Gy during the first five days. In the middle of the course, starting from the fifteenth day, four days of absence of irradiation are situated, which are preceded by one strong and one mild irradiation of about 3.6 and 1.5 Gy correspondingly, and are followed by four days of intermittent schedule, in which each of the two irradiations of more than 3 Gy is paired by a following day of rest. The end of the course is marked by three fractions of increasing dose, the latter one achieving almost 6 Gy. However, it should be noted, that such dosage for fractionated RT can be excessive from a biological point of view.

These results of an implementation of the developed algorithm serve as its approbation rather than a ready proposal of optimization of investigated type of combined anti-tumor treatment, due to the following reasons. Firstly, the major disadvantage of similar algorithms is the fact that they are able to indicate only locally optimal solution. In attempt to find the globally optimal solution, we are planning to introduce the technique of simulated annealing within the scope of the algorithm, and to start its work with various initial RT schedules and various values of its parameters, δ and ϵ. However, it should be noted, that it will nevertheless be practically impossible to prove that the most optimal of the proposed schemes will indeed be the globally optimal. Secondly, we assume that the considered combined treatment can be further optimized by one promising modification of an algorithm, that is the possibility of the shift of the beginning of AAT, by which the escalation of hypoxia, that inhibits the RT efficiency in the end of its course, can be delayed. Thirdly and most importantly, investigation of the simulation of the tumor growth under only a single set of parameter values is of little use by itself. Although some qualitative recommendations can be derived from the presented optimized scheme – i.e., enhancement of the first and the last fractional doses and intermittent schedule in the middle of the course – it is unclear, which of them, and what new recommendations, will show up in the simulations under other values of model parameters. The solutions to the posed issues can not be quickly obtained due to the bulkiness of calculations, which is caused by the complexity of the tumor growth model, – e.g., the above discussed simulation, being performed in parallel on eight cores, took about a day on an average home PC.

Certainly, we understand that the results of mathematical modeling can be regarded as only preliminary ones. However, if the effectiveness of some qualitative recommendations for optimizing the RT fractionation will be demonstrated within a wide range of both tumor-specific and patient-specific model parameters, then, in our opinion, it will serve as a good reason for the necessity of their experimental verification.

Acknowledgements

This study was funded by RFBR according to the research project No. 19-01-00768. Numerical simulations were prepared with the support of the "RUDN University Program 5-100".

References

[1] J. Chapman, D. Dugle, A. Reuvers, B. Meeker, J. Borsa, International Journal of Radiation Biology and Related Studies in Physics, Chemistry and Medicine **26**, 383 (1974)
[2] R.P. Dings, M. Loren, H. Heun, E. McNiel, A.W. Griffioen, K.H. Mayo, R.J. Griffin, Clinical Cancer Research **13**, 3395 (2007)
[3] R.K. Jain, Science **307**, 58 (2005)
[4] M. Pàez-Ribes, E. Allen, J. Hudock, T. Takeda, H. Okuyama, F. Viñals, M. Inoue, G. Bergers, D. Hanahan, O. Casanovas, Cancer cell **15**, 220 (2009)
[5] R. Mazeron, B. Anderson, S. Supiot, F. Paris, E. Deutsch, Cancer treatment reviews **37**, 476 (2011)
[6] D.H. Schanne, A.L. Grosu, D.G. Duda, *Anti-angiogenics and Radiation Therapy* (Springer International Publishing, Cham, 2017), pp. 1–10
[7] M.B. Kuznetsov, V.V. Gubernov, A.V. Kolobov, Russian Journal of Numerical Analysis and Mathematical Modelling **33**, 225 (2018)
[8] K. Borkenstein, S. Levegrün, P. Peschke, Radiation research **162**, 71 (2004)
[9] A. Gago-Arias, P. Aguiar, I. Espinoza, B. Sánchez-Nieto, J. Pardo-Montero, Physics in Medicine & Biology **61**, 1204 (2016)
[10] A. Ergun, K. Camphausen, L.M. Wein, Bulletin of mathematical biology **65**, 407 (2003)
[11] Å. Søvik, E. Malinen, Ø.S. Bruland, S.M. Bentzen, D.R. Olsen, Physics in Medicine & Biology **52**, 499 (2006)
[12] K.H. Küfer, A. Scherrer, M. Monz, F. Alonso, H. Trinkaus, T. Bortfeld, C. Thieke, OR spectrum **25**, 223 (2003)
[13] M.M. Hadjiandreou, G.D. Mitsis, IEEE Transactions on Biomedical Engineering **61**, 415 (2013)
[14] A. Bratus, Y. Todorov, I. Yegorov, D. Yurchenko, Journal of Optimization Theory and Applications **159**, 590 (2013)
[15] J. Holash, P. Maisonpierre, D. Compton, P. Boland, C. Alexander, D. Zagzag, G. Yancopoulos, S. Wiegand, Science **284**, 1994 (1999)
[16] D. Shweiki, M. Neeman, A. Itin, E. Keshet, Proceedings of the National Academy of Sciences **92**, 768 (1995)
[17] B.M. Fu, S. Shen, American Journal of Physiology-Heart and Circulatory Physiology **284**, H2124 (2003)
[18] J.F. Fowler, The British journal of radiology **62**, 679 (1989)
[19] B.G. Wouters, J.M. Brown, Radiation research **147**, 541 (1997)

Viruses competition in the genotype space

Cristina Leon[1,*], *Vladimir* Popov[1,**], and *Vitaly* Volpert[1,2,3,***]

[1]Peoples Friendship University of Russia (RUDN University), 6 Miklukho-Maklaya St, Moscow, 117198, Russian Federation
[2]Institut Camille Jordan, UMR 5208 CNRS, University Lyon 1, 69622 Villeurbanne, France
[3]INRIA Team Dracula, INRIA Lyon La Doua, 69603 Villeurbanne, France

Abstract.
This paper is devoted to the study of persistence and evolution of two viruses taking into account virus mutation, reproduction, and genotype dependent mortality, either natural or determined by an antiviral treatment. The model describes the virus density distribution $u(x;t)$ for the first virus and $v(y;t)$ for the second one as functions of genotypes x and y considered as continuous variables and of time t. The model consists of a system of reaction-diffusion equations with integral terms characterizing virus competition for host cells. The analysis of the model shows the conditions of the existence of virus strains.

1 Introduction

Viruses are in a constant evolution owing to variation of their genetic structure as the result of the interaction of the replication, recombination or mutation [1]. Further, these changes in the virus genetic structure contribute to the appearance of new variations thereof. The variability of the viruses is considered to be one of the key factors in the pathogonesis of the respective infectious disease, besides the high genetic variety and quickness of mutation of some viruses makes the immune response ineffective and it can lead to the emergence of the resistance to the antiviral therapy [2]. A better understanding of infection development would allow us the improvement techniques of viral treatments and their effectiveness.

When a virus enters the host organism, it begins to contaminate uninfected cells starting its replication. The concentration of virus in the host organism can be affected by its elimination, either by the immune response, virus natural death or some antiviral treatment. In [3], a model that describes the evolution of virus density depending on the genotype is introduced and the conditions for the existence of virus strains are determined.

In this study we propose a mathematical model which considers the aforementioned processes that affects the concentration of virus in the host organism without taking into account the immune response, in the case of the existence of two viruses u, v in the host. We consider the system of equations:

*e-mail: merycris25@hotmail.com
**e-mail: popov-va@rudn.ru
***e-mail: volpert@math.univ-lyon1.fr

$$\begin{cases} \dfrac{\partial u}{\partial t} = D_1 \dfrac{\partial^2 u}{\partial x^2} + \alpha_1 u(1 - \beta_1 I(u) - \gamma_1 I(v)) - \sigma_1(x)u, \\ \\ \dfrac{\partial v}{\partial t} = D_2 \dfrac{\partial^2 v}{\partial y^2} + \alpha_2 v(1 - \beta_2 I(u) - \gamma_2 I(v)) - \sigma_2(y)v. \end{cases} \tag{1}$$

Theses equations describe the evolution of virus densities depending on the genotypes x and y respectively, considered as continuous variables and on time. Here D_1, D_2 are the coefficients of diffusion and parameters $\alpha_{1,2}, \beta_{1,2}, \gamma_{1,2}$ are positive constants. The first terms in the right-hand side of these equations characterize virus mutation, the second terms its reproduction, and the last terms the virus death. We now detail each of these terms of the first equation. They are similar in for the second equation.

- Assuming that there is a sequence of reversible mutations with consecutive genotypes x_i, we can write the equation for the density u_i of virus with genotype x_i:

$$\frac{du_i}{dt} = \mu(u_{i-1} - u_i) + \mu(u_{i+1} - u_i),$$

  where μ is the frequency of mutations. This equation represents a discretization of the diffusion equation with the diffusion coefficient proportional to μ.
- Virus multiplication term is proportional to the virus density u and to the quantity of uninfected host cells $(1 - \beta_1 I(u) - \gamma_1 I(v))$ which is valid for the acute stage of infection (virus multiplication phase). Here 1 is a dimensionless total number of cells, $\beta_1 I(u)$ and $\gamma_1 I(v)$ are the number of infected cells by virus u and v, respectively, which are proportional to the total viruses quantity:

$$I(u) = \int_{-\infty}^{\infty} u(x,t)dx \quad \text{and} \quad I(v) = \int_{-\infty}^{\infty} v(y,t)dy.$$

- The last term in the right-hand side of equation (1) describes virus natural death or its elimination by some antiviral treatment. Let us note that the death rate can depend on virus genotype x or y.

We consider the virus strain as density distribution concentrated around some genotype value. It is a non-negative solution of system (1) that decays at infinity. We will determine the conditions of the existence of such stationary solutions. The detailed analysis of the reaction-diffusion type equations as well as the principles of formulating a mathematical model of virus infections and immunology are presented in [4] and [5].

2 Existence of stationary solutions

Stationary solutions of system (1) considered on the whole axis satisfy the following system of equations:

$$\begin{cases} D_1 u'' + \alpha_1 u(1 - \beta_1 I(u) - \gamma_1 I(v)) - \sigma_1(x)u = 0, \\ D_2 v'' + \alpha_2 v(1 - \beta_2 I(u) - \gamma_2 I(v)) - \sigma_2(y)v = 0, \end{cases} \tag{2}$$

where

$$I(u) = \int_{-\infty}^{\infty} u(x)dx, \qquad I(v) = \int_{-\infty}^{\infty} v(y)dy.$$

In order to find an analytical solution of this system, consider piece-wise constant functions $\sigma_i(x)$:

$$\sigma_1(x) = \begin{cases} \sigma_1, & \text{when} \quad |x| \geq x_1 \\ 0, & \text{when} \quad |x| < x_1 \end{cases}, \qquad \sigma_2(y) = \begin{cases} \sigma_2, & \text{when} \quad |y| \geq y_1 \\ 0, & \text{when} \quad |y| < y_1 \end{cases},$$

$\sigma_{1,2} > 1$. We look for a non-negative bounded solution of system of equations (2).

We suppose for simplicity that $\alpha_1 = \alpha_2, \beta_1 = \beta_2, \gamma_1 = \gamma_2$, and we omit the subscript in what follows. Clearly, non-negative bounded solutions of equations (2) can exist only if $\beta I(u) + \gamma I(v) < 1$. Set

$$\alpha(1 - \beta I(u) - \gamma I(v)) = k^2. \tag{3}$$

Then we can rewrite system (2) as follows:

$$\begin{cases} D_1 u'' + k^2 u - \sigma_1(x) u = 0, \\ D_2 v'' + k^2 v - \sigma_2(y) v = 0. \end{cases}$$

We look for its solutions in the form:

$$\begin{cases} u(x) = c_2 e^{\lambda_1 x}, & x \leq -x_1 \\ u(x) = c_1 \cos(t_1 x), & |x| < x_1 \\ u(x) = c_2 e^{-\lambda_1 x}, & x \geq x_1 \\ v(y) = c_4 e^{\lambda_2 y}, & y \leq -y_1 \\ v(y) = c_3 \cos(t_2 y), & |y| < y_1 \\ v(y) = c_4 e^{-\lambda_2 y}, & y \geq y_1 \end{cases},$$

where c_1, c_2, c_3 and c_4 are some non-negative constants and

$$t_1 = \frac{k}{\sqrt{D_1}}, \quad t_2 = \frac{k}{\sqrt{D_2}}, \quad \lambda_1 = \sqrt{\frac{\sigma_1 - k^2}{D_1}}, \quad \lambda_2 = \sqrt{\frac{\sigma_2 - k^2}{D_2}}.$$

From the continuity of the solution and its first derivative at $x = \pm x_1$ and $y = \pm y_1$, we obtain the following equalities:

$$\begin{aligned} c_1 \cos(t_1 x_1) &= c_2 e^{-\lambda_1 x_1}, & c_1 t_1 \sin(t_1 x_1) &= c_2 \lambda_1 e^{-\lambda_1 x_1}, \\ c_3 \cos(t_2 y_1) &= c_4 e^{-\lambda_2 y_1}, & c_3 t_2 \sin(t_2 y_1) &= c_4 \lambda_2 e^{-\lambda_2 y_1}. \end{aligned} \tag{4}$$

Dividing the equations at right by the equations at left, we get equations with respect to k:

$$\begin{cases} k = \dfrac{\sqrt{\sigma_1 - k^2}}{\tan(k x_1^*)}, \\ \\ k = \dfrac{\sqrt{\sigma_2 - k^2}}{\tan(k y_1^*)}. \end{cases} \tag{5}$$

where $x_1^* = \frac{x_1}{\sqrt{D_1}}$ and $y_1^* = \frac{y_1}{\sqrt{D_2}}$. From equality (3) we can see that $\alpha = \alpha(\beta I(u) + \gamma I(v)) + k^2$, where $I(u), I(v) > 0$ and α, β, γ are positive constants. Therefore $k^2 < \alpha$ and we look for a solution $k < \sqrt{\alpha}$ of this equation.

We can identify three cases:

1. Let $u(x) \neq 0, v(y) = 0$. Then the value of k can be found from the first equality of system (5).

2. Let $u(x) = 0, v(y) \neq 0$. Then the value of k can be found from the second equality of system (5).

   Let us recall that we look for a solution $k < \sqrt{\alpha}$. In the case where only one component of the solution is different from 0, such solution exists if $x_1^* > \xi_1$ (in the first case) or $y_1^* > \xi_2$ (in the second case) is greater than the critical value

$$\xi_{1,2} = \frac{1}{\sqrt{\alpha}} \arctan\left(\sqrt{\frac{\sigma_{1,2}}{\alpha} - 1}\right),$$

   and it does exist if $x_1^* \leq \xi_1$ or $y_1^* < \xi_2$, respectively in each case.

3. Let $u(x) \neq 0, v(y) \neq 0$. In this case the value $k(< \sqrt{\alpha})$ can be found from the first equality in (5). This value of k inserted in the second equality in (5) allows us to determine the relation between y_1^* and σ_2 for which there exists a solution (Figure 1, left). In the case where σ_1 and σ_2 are given, we can determine the relation between y_1^* and x_1^* from equations of system (5). If $\sigma_1 = \sigma_2$, then $y_1^* = x_1^*$ (black line in Figure 1, right). If $\sigma_1 < \sigma_2$, then $y_1^* > x_1^*$ and such relation exists only if $\sigma_1 > k^2$. If $\sigma_1 > \sigma_2$, then $y_1^* < x_1^*$ and such relation exists only if $\sigma_2 > k^2$.

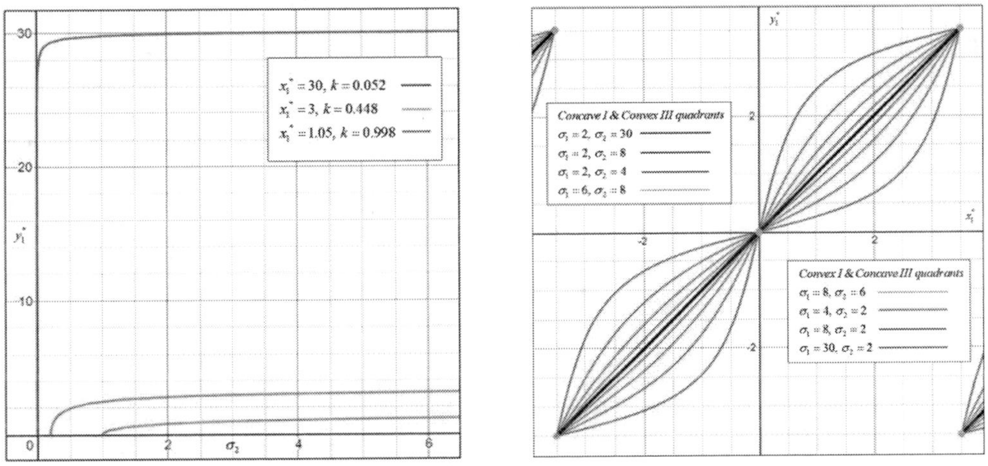

Figure 1. Left: $y_1^*(\sigma_2)$: constant values $\alpha = 1, \sigma_1 = 4$. Right: $y_1^*(x_1^*)$: *Black line* $\sigma_1 = \sigma_2$. *Concave part of the I and convex part of III quadrant* $\sigma_1 < \sigma_2$: $k = 0.448$. *Convex part of the I and concave part of III quadrant* $\sigma_1 > \sigma_2$: $k = 0.448$.

We can now determine the integrals $I(u)$ and $I(v)$:

$$I(u) = \int_{-\infty}^{\infty} u(x)dx = \frac{2c_1}{t_1}\sin(t_1 x_1) + \frac{2c_2}{\lambda_1}e^{-\lambda_1 x_1} \tag{6}$$

and

$$I(v) = \int_{-\infty}^{\infty} v(y)dy = \frac{2c_3}{t_2}\sin(t_2 y_1) + \frac{2c_4}{\lambda_2}e^{-\lambda_2 y_1}. \tag{7}$$

From (6) and (7), taking into account the first relations in (4), we have:

$$I(u) = 2c_1\left(\frac{1}{t_1}\sin(t_1x_1) + \frac{1}{\lambda_1}\cos(t_1x_1)\right),$$

$$I(v) = 2c_3\left(\frac{1}{t_2}\sin(t_2y_1) + \frac{1}{\lambda_2}\cos(t_2y_1)\right).$$

The coeffcients c_1 and c_3 can be determined from equation (3):

$$\begin{aligned} c_1 &= \frac{\alpha - \alpha\gamma I(v) - k^2}{2\alpha\beta h(k)}, \\ c_3 &= \frac{\alpha - \alpha\beta I(u) - k^2}{2\alpha\gamma g(k)}, \end{aligned} \tag{8}$$

where

$$h(k) = \sqrt{D_1}\left(\frac{1}{k}\sin\left(\frac{k}{\sqrt{D_1}}x_1\right) + \frac{1}{\sqrt{\sigma_1 - k^2}}\cos\left(\frac{k}{\sqrt{D_1}}x_1\right)\right),$$

$$g(k) = \sqrt{D_2}\left(\frac{1}{k}\sin\left(\frac{k}{\sqrt{D_2}}y_1\right) + \frac{1}{\sqrt{\sigma_2 - k^2}}\cos\left(\frac{k}{\sqrt{D_2}}y_1\right)\right).$$

We can solve linear system of equations (8) with respect to c_1 and c_3. Similarly, we can find the coefficients c_2 and c_4. An example of the analytical solution is shown in Figure 2. We see that $u(x)$ is positive and $v(y)$ is negative. We have not found the values of parameters for which both components of the soltuion are positives.

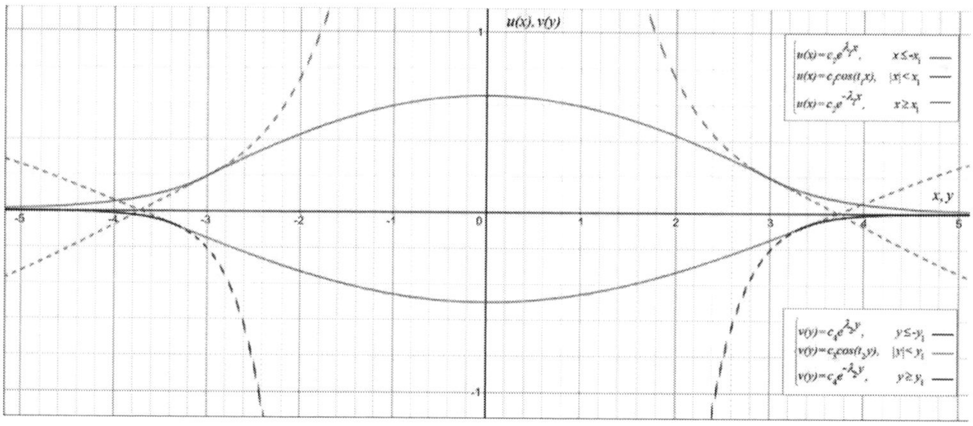

Figure 2. Graphical solution of system (2) for the values of parameters: $\alpha = \beta = \gamma = D_1 = D_2 = 1$, $\sigma_1 = 2$, $\sigma_2 = 8$, $x_1 = 3$, $y_1 = 3.3630$, $k = 0422$, $c_1 = 0.6478$, $c_2 = 11.0920$, $c_3 = -0.4995$, $c_4 = -906.7560$. The function $u(x)$ is described by the blue and red solid lines that smoothly continue each other at the point x_1. Similarly, the function $v(y)$ is composed by solid black and green lines.

3 Numerical simulations of system (1)

For numerical simulations, system (1) is considered in the interval $[0; L]$ with Neumann boundary conditions. The system is discretized by using an explicit finite differences scheme. The functions $\sigma_1(x)$ and $\sigma_2(y)$ are piece-wise constant. The integral is approximated by the trapeze method. Figure 3 shows the behavior of the solution of system (1) for the values of parameters $L = 12, \alpha = \beta = \gamma = D_1 = D_2 = 1; \sigma_1(x) = 0$ for $3 < x < 9$ and $= 2$ otherwise, initial condition $= 1$ for $5 < x < 7$; $\sigma_2(y) = 0$ for $2.637 < x < 9.363$ and $= 8$ otherwise, initial condition $= 1$ for $5.5 < x < 6.5$.

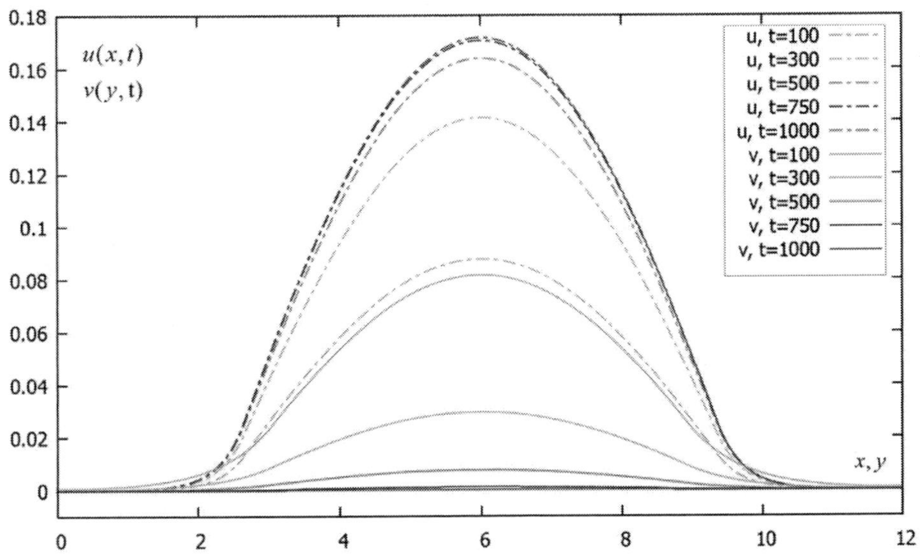

Figure 3. Solution $u(x, t)$ and $v(y, t)$ of system (1) in numerical simulations.

The behavior of this solution is characterized by a slow convergence to a stationary solution for which $u(x) > 0$ and $v(y) = 0$. The analytical solution considered in the previous section shows the existence of a stationary solution with a positive component $u(x)$ and a negative component $v(y)$. However, since the initial condition is positive for both components, the solution of system (1) remains also positive and converges to another stationary solution having one positive component and one zero component.

4 Discussion

In this work, we study the competition of two viruses in the host organism. The model takes into account virus mutation, reproduction, and mortality. The nonlocal terms describe virus multiplication and competition for uninfected host cells. Virus mortality depends on the genotype and it can be due to antiviral treatment or to its natural death.

We consider virus strain as a density distribution concentrated around some genotype value. This is a non-negative solution decaying at infinity. The analysis of the model shows the existence of such solutions under some conditions on parameters. Namely, the admissible interval where virus multiplication rate is larger than its mortality rate should be sufficiently long, and the mutation rate should be small enough.

We identify three types of nonzero solutions: in two of them, one component of the solution is positive, and another one is zero. There is one more solution with one positive and one negative components. This solution does not have biological meaning, and it is not considered in the applications. Thus, the competition between two viruses results in the disappearance of one of them and persistence of another one. We have not found a solution with two positive components. This can be due to the assumptions of the equality of the coefficients $\alpha_1 = \alpha_2, \beta_1 = \beta_2, \gamma_1 = \gamma_2$. More general case where these coefficients are different will be considered in the future works.

Let us note that variation of the death function in time allows the investigation of the influence of an antiviral treatment and virus evasion. In the case of one equation model this question was studied in [3]. For the two equation model considered in this work this question remains open.

References

[1] E. Domingo, C. Perales, Eur Biophys J. **47**, 443-457 (2018).

[2] J. Coffin, R. Swanstrom, Cold Spring Harb Perspect Med. **3(1)**:a012526 (2013).

[3] N. Bessonov, G. Bocharov, A. Meyerhans, V. Popov, V. Volpert, Mathematics **8(1)**, 117 (2020).

[4] V. Volpert, *Elliptic partial differential equations* (Birkhäuser Basel, 2014) 784.

[5] G. Bocharov, V. Volpert, B. Ludewig, A. Meyerhans, *Mathematical Immunology of Virus Infections* (Springer International Publishing, 2018) 245.

ITM Web of Conferences **31**, 02003 (2020) https://doi.org/10.1051/itmconf/20203102003
Mathematical Modelling in Biomedicine 2019

Stochastic compartmental model of HIV-1 infection

Konstantin Loginov[1,*] and *Nikolai* Pertsev[1,**]

[1]Sobolev Institute of Mathematics, Siberian Branch of RAS, Omsk Department; 13 Pevtsova avenue, 644043, Omsk, Russia.

Abstract. Stochastic model of the dynamics of HIV-1 infection describing the interaction of target cells and viral particles in the lymphatic nodes and their movement between the lymphatic nodes is constructed. The lymphatic system is represented as a graph, vertices of which are the lymphatic nodes and edges are the lymphatic vessels. The novelty of the model consists in the description of populations of cells and viral particles in terms of a multidimensional birth and death process with the random point-distributions. The random point-distributions describe the duration of the transition of cells and viral particles between the lymph nodes and the duration of the stages of their development. The durations of transitions of viral particles and cells between the lymphatic nodes are not random and based on the rate of lymph flow. The durations of the developmental stages of infected target cells are assume to be constant. The graph theory for the formalization and compact representation of the model is used. An algorithm for modelling the dynamics of the studied populations is constructed basing on the Monte-Carlo method. The results of computational experiments for a system consisting of five lymphatic nodes are presented.

1 Introduction

An important direction in mathematical modeling in immunology is the development of mathematical models of the dynamics of HIV-1 infection in the human lymphatic system [1, 2]. Mostly, these models are divided into deterministic and stochastic one. Deterministic models take the form of the systems of delay differential equations [3, 4]. The equations of such models describe concentration of target cells, infected cells, viral particles, etc. An example of a stochastic model is constructed at [5] in the form of multidimensional random birth and death process. A stochastic model with integer variables describing the dynamics of HIV-1 infection in a single lymph node in the form of random process with particle interaction is proposed at [6]. The lymphatic system is a collection of lymph nodes interconnected by lymphatic vessels through which cells and viral particles can move. Modelling movements of cells in the lymphatic system is needed the construction of a stochastic compartment model with pipes or its deterministic analogue. One of the first deterministic models in this direction presented at [7]. An important feature of the model described at [7] is the presence of transport delays, which reflect the duration of particle movement through the pipes. Other models of the lymphatic system based on partial differential equations including the Navier-Stokes equations are presented, for example at [8]. Mathematical model of the dynamics of

*e-mail: kloginov85@mail.ru
**e-mail: homlab@ya.ru

HIV-1 infection in the form of high-dimensional system of ordinary differential equations is presented at [9]. The model described at [9] contains detailed information regarding the structure of the human lymphatic system, however, there are no transfer delays in the model equations. It is important to note that the absence of transport delays leads to the possibility of instantaneous and simultaneous changes in the number of populations in various lymph nodes caused by processes occurring in only one specific lymph node. Stochastic model with one type of particles moving in a compartment system with pipes is considered at [10].

The purpose of this article is to build a stochastic compartmental model of the dynamics of HIV-1 infection which combines the approaches proposed at [6, 10].

2 Assumptions on the dynamics of HIV-1 infection

In describing the dynamics of HIV-1 infection, we will use the following notation:

- T denotes T-lymphocytes (target cells for mature viral particles);
- C is infected cells, i.e. cells in preparation for the production of viral particles;
- I is productively infected cells (cells producing viral particles);
- U denotes immature viral particles;
- V corresponds to mature viral particles (or virions);
- K is precursor cells of effector lymphocytes;
- E is related to cytotoxic T-lymphocytes (effectors).

Moreover, let S denotes the cell source in bone marrow and lymphoid tissue generating the target cells T and cells K; D is all dead cells and viral particles.

We will describe the lymphatic system in the form of graph G. The notations are as follows:

- G is oriented graph without loops with weighted edges;
- n is number of graph vertices G;
- N_i denotes vertices of the graph G (lymph nodes), $1 \leqslant i \leqslant n$;
- N_{ij} corresponds edges of the graph G (lymph vessels), $1 \leqslant i, j \leqslant n$, $i \neq j$;
- $Q = \|q_{ij}\|$ is matrix defining relationships between vertices of the graph G; constants $0 \leqslant q_{ij} \leqslant 1$ for all $i = 1, ..., n$ satisfy the conditions:

$$q_{ii} = 0; \quad q_{ij} \cdot q_{ji} = 0, 1 \leqslant j \leqslant n, j \neq i; \quad \sum_{j=1}^{n} q_{ij} = 1.$$

  If $q_{ij} > 0$, then vertices N_i and N_j interconnected by edge N_{ij} but connection between N_j and N_i is absent. For $q_{ij} = 0$ there is no connection between N_i and N_j but there may be a connection between N_j and N_i.

- $\Delta_{ij}(t)$ is positive and bounded above real-valued function that defines the weight of the edge N_{ij}, $t \in (-\infty; \infty)$, $q_{ij} > 0$. We assume that $\Delta_{ij}(t) + t$ is strictly increasing.

Example of the graph G is presented on the figure 1.

Denote

- T_{ii}, C_{ii}, I_{ii}, U_{ii}, V_{ii}, K_{ii}, E_{ii} are components of the immune system and HIV-1 particles located in the vertex N_i, $1 \leqslant i \leqslant n$;

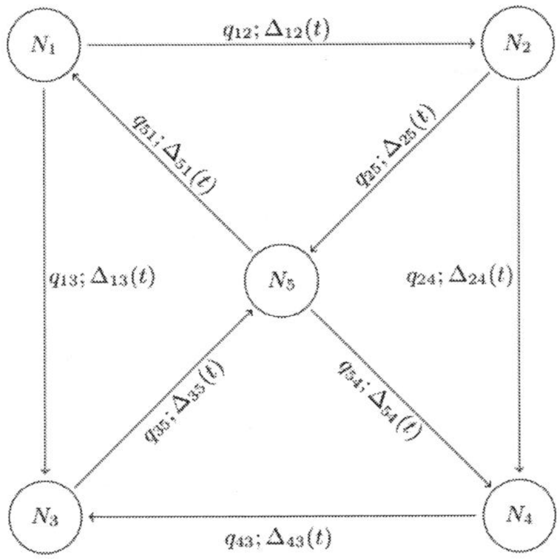

Figure 1. Example of the graph G with five vertices

- T_{ij}, C_{ij}, I_{ij}, U_{ij}, V_{ij}, K_{ij}, E_{ij} are components of the immune system and HIV-1 particles in the state of transition from vertex N_i to vertex N_j along edge N_{ij}, $1 \leqslant i, j \leqslant n, i \neq j$.

Following [2, 4, 6, 10], we introduce the set of assumptions that determine dynamics of HIV-1 infection:

- Inflow of target cells T to vertex N_i with rate $r_{ii}^{(T)} > 0$: $S \xrightarrow{r_{ii}^{(T)}} S + T_{ii}$.
- Natural death of cells T, C, I, E and particles U, V:

$$T_{ii} \xrightarrow{\mu_{ii}^{(T)}} D;\ C_{ii} \xrightarrow{\mu_{ii}^{(C)}} D;\ I_{ii} \xrightarrow{\mu_{ii}^{(I)}} D;\ E_{ii} \xrightarrow{\mu_{ii}^{(E)}} D;$$
$$U_{ii} \xrightarrow{\mu_{ii}^{(U)}} D;\ V_{ii} \xrightarrow{\mu_{ii}^{(V)}} D.$$

- Interaction of T-cells with virions V (cell infection) with rate $\gamma_{ii}^{(T,V)} > 0$ per one pair of (T, V): $T_{ii} + V_{ii} \xrightarrow{\gamma_{ii}^{(T,V)}} C_{ii}$.
- Infection of cells T as a result of interaction with productively infected cells I with rate $\gamma_{ii}^{(T,I)} > 0$ per one pair of (T, I): $T_{ii} + I_{ii} \xrightarrow{\gamma_{ii}^{(T,I)}} C_{ii} + I_{ii}$.
- Production by cells I of immature viral particles of U with rate $\nu_{ii}^{(U)} > 0$: $I_{ii} \xrightarrow{\nu_{ii}^{(U)}} I_{ii} + U_{ii}$.
- Destruction of infected cells I by cytotoxic cells E with rate $\gamma_{ii}^{(I,E)} > 0$ per one pair of (I, E): $I_{ii} + E_{ii} \xrightarrow{\gamma_{ii}^{(I,E)}} D + E_{ii}$.
- Cell death I due to the destructive process of production of viral particles with rate $\sigma_U\, \nu_{ii}^{(U)}$, $\sigma_U > 0$: $I_{ii} \xrightarrow{\sigma_U\, \nu_{ii}^{(U)}} D$.
- Formation of precursor cells K cytotoxic lymphocytes E with rate $\nu_{ii}^{(K)} > 0$: $I_{ii} \xrightarrow{\nu_{ii}^{(K)}} I_{ii} + K_{ii}$.
- The cell C which appeared at time t and did not die during the time interval $(t, t + \omega_C)$, $\omega_C > 0$, at the time $t + \omega_C$ transforms into a productively infected cell I: $C_{ii}|_{\omega_C} \longrightarrow I_{ii}$.

- The immature virus particle U that arose at time t and did not die during the time interval $(t, t + \omega_U)$, $\omega_U > 0$, at the time $t + \omega_U$ transforms into virion V: $U_{ii}|_{\omega_U} \longrightarrow V_{ii}$.
- The cell K which arose at time t, at time $t + \omega_K$, $\omega_K > 0$, transforms into a collection of $n_E > 0$ cytotoxic T-lymphocytes E: $K_{ii}|_{\omega_K} \longrightarrow n_E\, E_{ii}$.
- Transformation of the cells T_{ii}, I_{ii}, E_{ii} and virions V_{ii} into T_{ij}, I_{ij}, E_{ij}, V_{ij} (start of cell and virion transfer) with rates $\beta_{ij}^{(T)} = q_{ij}^{(T)} \lambda_{ii}^{(T)}$, $\beta_{ij}^{(I)} = q_{ij}^{(I)} \lambda_{ii}^{(I)}$, $\beta_{ij}^{(E)} = q_{ij}^{(E)} \lambda_{ii}^{(E)}$, $\beta_{ij}^{(V)} = q_{ij}^{(V)} \lambda_{ii}^{(V)}$ respectively:

$$T_{ii} \xrightarrow{\beta_{ij}^{(T)}} T_{ij}; \quad I_{ii} \xrightarrow{\beta_{ij}^{(I)}} I_{ij}; \quad E_{ii} \xrightarrow{\beta_{ij}^{(E)}} E_{ij}; \quad V_{ii} \xrightarrow{\beta_{ij}^{(V)}} V_{ij};$$

where $q_{ij}^{(\cdot)} \geqslant 0$ is probability that the transition of cells and virions will occur from vertex N_i to vertex N_j along edge N_{ij}; $\lambda_{ii}^{(\cdot)} > 0$ is rate at which cells and virions begin to transition from vertex N_i to vertex N_j along edge N_{ij}; $1 \leqslant i, j \leqslant n$, $i \neq j$.

The assumptions made are shown in figure 2.

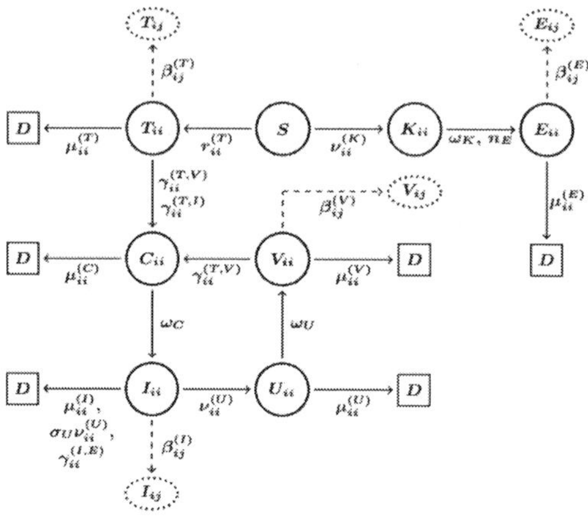

Figure 2. Scheme of the model for a single lymph node N_i

Let us introduce a set of assumptions that determine the movement of cells T, I, E and virions V along the edges of the graph G:

- The cells C, K and viral particles U don't move along the edges (lymphatic vessels) of the graph G.
- Interactions of cells and virions in edges of G is absent.
- Natural death of cells T_{ij}, I_{ij}, E_{ij} and virions V_{ij}:

$$T_{ij} \xrightarrow{\mu_{ij}^{(T)}} D; \quad I_{ij} \xrightarrow{\mu_{ij}^{(I)}} D; \quad E_{ij} \xrightarrow{\mu_{ij}^{(E)}} D; \quad V_{ij} \xrightarrow{\mu_{ij}^{(V)}} D.$$

- The cells T_{ij}, I_{ij}, E_{ij} and virions V_{ij} that started transition from vertex N_i to vertex N_j along edge N_{ij} at the moment t and didn't die during the transition, at the time

$$\eta_{ij}^{(T)}(t) = t + \Delta_{ij}^{(T)}(t); \quad \eta_{ij}^{(I)}(t) = t + \Delta_{ij}^{(I)}(t);$$
$$\eta_{ij}^{(E)}(t) = t + \Delta_{ij}^{(E)}(t); \quad \eta_{ij}^{(V)}(t) = t + \Delta_{ij}^{(V)}(t);$$

reaches the vertex N_j and transforms into T_{jj}, I_{jj}, E_{jj}, V_{jj} respectively.

The above assumptions have the following probabilistic interpretation. Fix the indexes $1 \leqslant i, j \leqslant n$, $i \neq j$. Assume, that $h \to +0$.

The expression $S \xrightarrow{r_{ii}^{(T)}} S + T_{ii}$ means that over a period of time $(t; t+h)$, with probability $r_{ii}^{(T)} h + o(h)$ to the vertex N_i from S arrives one cell T_{ii}, and with probability $1 - r_{ii}^{(T)} h + o(h)$ cell entry does not occur.

The expression $T_{ii} \xrightarrow{\mu_{ii}^{(T)}} D$ means that over a period of time $(t; t+h)$, with probability $\mu_{ii}^{(T)} h + o(h)$ one cell T_{ii} dies, and with probability $1 - \mu_{ii}^{(T)} h + o(h)$ does not die.

The expression $T_{ii} + V_{ii} \xrightarrow{\gamma_{ii}^{(T,V)}} C_{ii}$ means that over a period of time $(t; t+h)$, with probability $\gamma_{ii}^{(T,V)} h + o(h)$ one T_{ii} cell interacts with one virion V and with probability $1 - \gamma_{ii}^{(T,V)} h + o(h)$ such interaction does not occur.

The expression $T_{ii} \xrightarrow{\beta_{ij}^{(T)}} T_{ij}$ means that over a period of time $(t; t+h)$, with probability $\beta_{ij}^{(T)} h + o(h)$ one T_{ii} cell turns into one T_{ij} cell, and with probability $1 - \beta_{ij}^{(T)} h + o(h)$ does not turn.

The expression $T_{ij} \xrightarrow{\mu_{ij}^{(T)}} D$ means that over a period of time $(t; t+h)$, with probability $\mu_{ij}^{(T)} h + o(h)$ one cell T_{ij} dies, and with probability $1 - \mu_{ij}^{(T)} h + o(h)$ does not die.

Other expressions have a similar meaning. Denote

$$X_{ii}(t) = (T_{ii}(t), C_{ii}(t), I_{ii}(t), U_{ii}(t), V_{ii}(t), K_{ii}(t), E_{ii}(t)),$$
$$X_{ij}(t) = (T_{ij}(t), C_{ij}(t), I_{ij}(t), U_{ij}(t), V_{ij}(t), K_{ij}(t), E_{ij}(t))$$

vectors containing the number of system components at the vertex N_i and at the edge N_{ij} respectively (for each fixed t the components of the vectors are non-negative integer random variables). Note that in this model $C_{ij}(t) = U_{ij}(t) = K_{ij}(t) = 0$ for each $t \geqslant 0$, $i \neq j$. To account for the stages of development of cells and viral particles, we introduce random point-distributions. Random point-distributions contain information about the moments of the appearance of maturing cells and particles, as well as the moments of their transformation into the corresponding cells and particles. For instance, random point-distribution, using to describe C_{ii} cells, has the structure

$$\Omega_{ii}^{(C)}(t) = \{\tau_{ii}^{(C1)} + \omega_C, \ldots, \tau_{ii}^{(Ck)} + \omega_C, \ldots, \tau_{ii}^{(CC_{ii}(t))} + \omega_C\},$$
$$\tau_{ii}^{(C1)} < \cdots < \tau_{ii}^{(Ck)} < \cdots < \tau_{ii}^{(CC_{ii}(t))} \leqslant t,$$
$$\tau_{ii}^{(Ck)} + \omega_C > t, \quad 1 \leqslant k \leqslant C_{ii}(t),\ 1 \leqslant i \leqslant n,$$

where $\tau_{ii}^{(Ck)}$ is the moment of the appearance of another infected cell C; $\tau_{ii}^{(Ck)} + \omega_C$ is the moment of transformation of an infected cell into a productively infected cell I (without taking into account death). If $C_{ii}(t) = 0$ for some t, then $\Omega_{ii}^{(C)}(t) = \emptyset$. For particles U_{ii} and cells K_{ii} random point-distributions are constructed similarly.

To describe a population of cells T_{ij}, I_{ij}, E_{ij} and virions V_{ij} we also use random point-distributions reflecting the movement of particles from vertex N_i to vertex N_j, $i \neq j$. For instance, random point-distribution, using to describe T_{ij} cells, has the structure

$$\Omega_{ij}^{(T)}(t) = \left\{\eta_{ij}^{(T)}(\tau_{ij,1}^{(T)}); \ldots; \eta_{ij}^{(T)}(\tau_{ij,k}^{(T)}); \ldots; \eta_{ij}^{(T)}(\tau_{ij,\zeta(t)}^{(T)})\right\},$$
$$\tau_{ij,1}^{(T)} < \cdots < \tau_{ij,\zeta(t)}^{(T)} \leqslant t < \eta_{ij}^{(T)}(\tau_{ij,1}^{(T)}) < \cdots < \eta_{ij}^{(T)}(\tau_{ij,\zeta(t)}^{(T)}),$$

where $\tau_{ij,k}^{(T)}$ is the moment of the beginning of the transition of a certain particle T_{ij} from vertex N_i to vertex N_j along the edge N_{ij}; and $\eta_{ij}^{(T)}(\tau_{ij,k}^{(T)}) = \tau_{ij,k}^{(T)} + \Delta_{ij}^{(T)}(\tau_{ij,k}^{(T)})$ is the moment of completion of the transition provided that the specified particle does not die during the transition. If $T_{ij}(t) = 0$ for some t, then $\Omega_{ij}^{(T)}(t) = \emptyset$. For I_{ij}, E_{ij} and V_{ij} random point-distributions are constructed similarly.

Next, we show the state change of the studied random process at discrete moments of time. Assume $t_0 = 0$; $X_{ij}(t_0) \geqslant 0$ are initial numbers of system components at vertices and edges of graph G; $\Omega_{ii}^{(C)}(t_0)$, $\Omega_{ii}^{(U)}(t_0)$, $\Omega_{ii}^{(K)}(t_0)$, $\Omega_{ij}^{(T)}(t_0)$, $\Omega_{ij}^{(I)}(t_0)$, $\Omega_{ij}^{(E)}(t_0)$, $\Omega_{ij}^{(V)}(t_0)$ are fixed random point-distributions, $1 \leqslant i, j \leqslant n$, $i \neq j$.

Fix t_m, $X_{ij}(t_m)$, $\Omega_{ii}^{(C)}(t_m)$, $\Omega_{ii}^{(U)}(t_m)$, $\Omega_{ii}^{(K)}(t_m)$, $\Omega_{ij}^{(T)}(t_m)$, $\Omega_{ij}^{(I)}(t_m)$, $\Omega_{ij}^{(E)}(t_m)$, $\Omega_{ij}^{(V)}(t_m)$ for some $m = 0, 1, 2, \ldots$ Let

$$
\begin{gathered}
t_{m+1} = \min\left\{\eta^{(m)}, t_m + \psi^{(m)}\right\}, \quad X_{ij}(t_{m+1}) = X_{ij}(t_m) + \Delta X_{ij}(t_m), \\
\Omega_{ii}^{(C)}(t_{m+1}) = \Omega_{ii}^{(C)}(t_m) \cup \{t_{m+1} + \omega_C\}, \\
\text{or } \Omega_{ii}^{(C)}(t_{m+1}) = \Omega_{ii}^{(C)}(t_m) \setminus \{\tau_{ii}^{(C1)} + \omega_C\} \\
(\text{for } \Omega_{ii}^{(U)}(t_{m+1}),\ \Omega_{ii}^{(K)}(t_{m+1}) \text{ similarly } \Omega_{ii}^{(C)}(t_{m+1})); \\
\Omega_{ij}^{(T)}(t_{m+1}) = \Omega_{ij}^{(T)}(t_m) \cup \{\eta_{ij}^{(T)}(t_{m+1})\}, \\
\text{or } \Omega_{ij}^{(T)}(t_{m+1}) = \Omega_{ij}^{(T)}(t_m) \setminus \{\eta_{ij}^{(T)}(\tau_{ij,1}^{(T)})\} \\
(\text{for } \Omega_{ij}^{(I)}(t_{m+1})),\ \Omega_{ij}^{(E)}(t_{m+1})),\ \Omega_{ij}^{(V)}(t_{m+1}) \text{ similarly } \Omega_{ij}^{(T)}(t_{m+1})).
\end{gathered}
$$

Random variable $\eta^{(m)}$ is the nearest to t_m time when any of cells T_{ij}, I_{ij}, E_{ij} or virions V_{ij} reach the vertex N_j along the edge N_{ij} among all vertices N_i and all edges N_{ij}:

$$
\begin{gathered}
\eta^{(m)} = \min\{\eta^{(m,T)}, \eta^{(m,I)}, \eta^{(m,E)}, \eta^{(m,V)}\}, \\
\eta^{(m,T)} = \min_{(i,j)\in(I,J)_{m,T}\neq\emptyset} \{\eta_{ij}^{(T)}(\tau_{ij,1}^{(T)})\}, \\
(I, J)_{m,T} = \{1 \leqslant i, j \leqslant n, i \neq j : T_{ij}(t_m) > 0\}.
\end{gathered}
$$

If $(I, J)_{m,T} = \emptyset$, assign $\eta^{(m,T)} = +\infty$. Similarly for $\eta^{(m,I)}$, $\eta^{(m,E)}$ and $\eta^{(m,V)}$.

Random variable

$$
t_m + \psi^{(m)} = \min_{1\leqslant i\leqslant n}\{\tau_{ii}^{(C1)} + \omega_C, \tau_{ii}^{(U1)} + \omega_U, \tau_{ii}^{(K1)} + \omega_K, t_m + \xi_{ii}(t_m)\},
$$

where $\tau_{ii}^{(C1)} + \omega_C$, $\tau_{ii}^{(U1)} + \omega_U$, $\tau_{ii}^{(K1)} + \omega_K$, are the minimal (first) elements of point-distributions $\Omega_{ii}^{(C)}(t_m)$, $\Omega_{ii}^{(U)}(t_m)$, $\Omega_{ii}^{(K)}(t_m)$ respectively; $\xi_{ii}(t_m)$ is random variable representing the duration of time until one of the events at N_i:

- inflow of cell T from S to the vertex N_i;
- death of cells or viral particles at the vertex N_i;
- interaction of cells and viral particles at the vertex N_i;
- beginning of the transition of cells and viral particles from vertex N_i to vertex N_j along the edge N_{ij}.

3 Special case of the model

Without taking into account the transfer of cells and virions, the stochastic model includes the trivial special case. Let $C_{ii}(0) = I_{ii}(0) = U_{ii}(0) = V_{ii}(0) = K_{ii}(0) = E_{ii}(0) = 0$, $1 \leqslant i \leqslant n$.

Denote by $\mathbf{E}\,X_{ii}(t) = (\mathbf{E}\,T_{ii}(t), \mathbf{E}\,C_{ii}(t), \mathbf{E}\,I_{ii}(t), \mathbf{E}\,U_{ii}(t), \mathbf{E}\,V_{ii}(t), \mathbf{E}\,K_{ii}(t), \mathbf{E}\,E_{ii}(t))$ mathematical expectation of the $X_{ii}(t)$. Using the postulates of the random birth and death process, we can derive the differential equation for the mathematical expectation $T^{(i)}(t) = \mathbf{E}\,T_{ii}(t)$:

$$\frac{dT^{(i)}(t)}{dt} = r_{ii}^{(T)} - \mu_{ii}^{(T)} T^{(i)}(t), \quad t \geqslant 0; \quad T^{(i)}(0) = \mathbf{E}\,T_{ii}(0).$$

The other components of vector $\mathbf{E}\,X_{ii}(t)$ are identically zero. One can see, that

$$T^{(i)}(t) \longrightarrow T_{ii}^{*} = r_{ii}^{(T)} / \mu_{ii}^{(T)}, \quad t \to +\infty.$$

Following [6], we introduce the quotient basic reproductive number for a single lymph node N_i:

$$R_i^{(0)} = \frac{\gamma_{ii}^{(T,I)} T_{ii}^{*} e^{-\mu_{ii}^{(C)} \omega_C}}{\mu_{ii}^{(I)} + \sigma_U \nu_{ii}^{(U)}} + \frac{\nu_{ii}^{(U)} \gamma_{ii}^{(T,V)} T_{ii}^{*} e^{-(\mu_{ii}^{(C)} \omega_C + \mu_{ii}^{(U)} \omega_U)}}{(\mu_{ii}^{(I)} + \sigma_U \nu_{ii}^{(U)})(\mu_{ii}^{(V)} + \gamma_{ii}^{(T,V)} T_{ii}^{*})}.$$

The results of computational experiments presented at [6] showed that for $R_i^{(0)} < 1$

$$\mathbf{E}\,X_{ii}(t) \longrightarrow (T_{ii}^{*}, 0, 0, 0, 0, 0, 0), \quad t \to +\infty.$$

4 Computational experiments

The purpose of computational experiments is to demonstrate the dynamics of mathematical expectations of variables depending on the variation of values of some model parameters. Number of vertices $n = 5$, structure of the graph G is presented in figure 1. The values of the model parameters were chosen so that $R_i^{(0)} < 1$ for all vertices or $R_i^{(0)} > 1$ for at least one of the vertices. The simulation was carried out using the Monte–Carlo method on the time interval $[0; W]$. The number of independent realizations of the studied random process is $N = 1000$. Components of mathematical expectation $\mathbf{E}\,X_{ii}(t)$ were estimated by standard formulas of mathematical statistics. We used point $\bar{X}_{ii}(t)$ and interval $\bar{X}_{ii}(t) \pm 1.96\,\sigma_{X_{ii}(t)}$ estimates at a confidence level $P = 0.95$. For each of two experiments the maximum value of $\sigma_{X_{ii}(t)}$ for the components $\mathbf{E}\,X_{ii}(t)$ does not exceed 1.62. For computational experiments we used a reference set of parameters, the numerical values of which were selected from the articles [2–

4, 8, 11, 12]:

$$
\begin{aligned}
&r_{11}^{(T)} = 103,\ r_{22}^{(T)} = 90,\ r_{33}^{(T)} = 95,\ r_{44}^{(T)} = 97,\ r_{55}^{(T)} = 80;\\
&\mu_{ij}^{(T)} = \mu_{ij}^{(I)} = 0.01, \quad 1 \leqslant i, j \leqslant 5;\\
&\mu_{ii}^{(C)} = 0.01, \quad \mu_{ij}^{(C)} = 0, \quad 1 \leqslant i, j \leqslant 5, \ i \neq j;\\
&\mu_{ii}^{(U)} = 3, \quad \mu_{ij}^{(U)} = 0, \quad 1 \leqslant i, j \leqslant 5, \ i \neq j;\\
&\mu_{ij}^{(V)} = 3, \quad \mu_{ij}^{(E)} = 0.43, \quad 1 \leqslant i, j \leqslant 5;\\
&\gamma_{11}^{(T,V)} = \gamma_{22}^{(T,V)} = \gamma_{33}^{(T,V)} = 6.5 \cdot 10^{-7},\ \gamma_{44}^{(T,V)} = 4 \cdot 10^{-7},\\
&\gamma_{55}^{(T,V)} = 5.5 \cdot 10^{-7},\ \gamma_{ij}^{(T,V)} = 0, \quad 1 \leqslant i, j \leqslant 5, \ i \neq j;\\
&\gamma_{11}^{(T,I)} = \gamma_{22}^{(T,I)} = \gamma_{44}^{(T,I)} = 5 \cdot 10^{-6},\ \gamma_{33}^{(T,I)} = 10^{-7},\\
&\gamma_{55}^{(T,I)} = 10^{-9},\ \gamma_{ij}^{(T,I)} = 0, \quad 1 \leqslant i, j \leqslant 5, \ i \neq j;\\
&\gamma_{ii}^{(I,E)} = 3.7 \cdot 10^{-4}, \quad \gamma_{ij}^{(I,E)} = 0, \quad 1 \leqslant i, j \leqslant 5, \ i \neq j;\\
&\nu_{11}^{(U)} = \nu_{22}^{(U)} = \nu_{33}^{(U)} = \nu_{44}^{(U)} = 110,\ \nu_{55}^{(U)} = 200,\\
&\nu_{ij}^{(U)} = 0, \quad 1 \leqslant i, j \leqslant 5, \ i \neq j;\\
&\nu_{ii}^{(K)} = 0.05, \quad \nu_{ij}^{(K)} = 0, \quad 1 \leqslant i, j \leqslant 5, \ i \neq j; \quad (\text{day}^{-1});\\
&\omega_C = 0.2, \quad \omega_U = 0.021, \quad \omega_K = 1; \quad (\text{day});\\
&\sigma_U = 2.618 \cdot 10^{-3}, \quad n_E = 10.
\end{aligned}
$$

Parameters of transfer of cells and virions between nodes:

$$
\begin{aligned}
&\lambda_{11}^{(T)} = \lambda_{11}^{(I)} = \lambda_{11}^{(V)} = \lambda_{11}^{(E)} = 0.25,\\
&\lambda_{22}^{(T)} = \lambda_{22}^{(I)} = \lambda_{22}^{(V)} = \lambda_{22}^{(E)} = 0.45,\\
&\lambda_{33}^{(T)} = \lambda_{33}^{(I)} = \lambda_{33}^{(V)} = \lambda_{33}^{(E)} = 0.08,\\
&\lambda_{44}^{(T)} = \lambda_{44}^{(I)} = \lambda_{44}^{(V)} = \lambda_{44}^{(E)} = 0.06,\\
&\lambda_{55}^{(T)} = \lambda_{55}^{(I)} = \lambda_{55}^{(V)} = \lambda_{55}^{(E)} = 0.37; \quad (\text{day}^{-1});
\end{aligned}
$$

$$
Q^{(T)} = \|q_{ij}^{(T)}\| = \begin{pmatrix} 0 & 0.4 & 0.6 & 0 & 0 \\ 0 & 0 & 0 & 0.444 & 0.556 \\ 0 & 0 & 0 & 0 & 1 \\ 0 & 0 & 1 & 0 & 0 \\ 0.676 & 0 & 0 & 0.324 & 0 \end{pmatrix};
$$

$$
Q^{(I)} = Q^{(V)} = Q^{(E)} = Q^{(T)};
$$

$$
\begin{aligned}
\Delta_{ij}^{(T)}(t) = 0.083 + 0.1\ \cos(t) + 0.2\ \cos(2\,t) + \\
+ 0.05\ \cos(3\,t) + 0.25\ \cos(0.5\,t); \quad (\text{day});\\
\Delta_{ij}^{(I)}(t) = \Delta_{ij}^{(V)}(t) = \Delta_{ij}^{(E)}(t) = \Delta_{ij}^{(T)}(t),\ 1 \leqslant i, j \leqslant 5,\ i \neq j.
\end{aligned}
$$

The initial conditions for all experiments:

$$
\begin{aligned}
C_{ij}(0) = I_{ij}(0) = U_{ij}(0) = K_{ij}(0) = E_{ij}(0) = 0,\\
T_{ii}(0) = r_{ii}^{(T)} / \mu_{ii}^{(T)}, \ \ T_{ij}(0) = 0, \ \ 1 \leqslant i, j \leqslant 5, \ i \neq j.
\end{aligned}
$$

Experiment 1. The reference set of parameters was used; the quotient basic reproductive numbers are as follows:

$$R_1^{(0)} = 0.94, \quad R_2^{(0)} = 0.82, \quad R_3^{(0)} = 0.71,$$
$$R_4^{(0)} = 0.61, \quad R_5^{(0)} = 0.51.$$

On the figures 3, 4 dynamics of the point estimates $T^{(i)}(t)$ without HIV-1 infection, i.e. $V_{ij}(0) = 0$, and with HIV-1 infection, i.e. $V_{ii}(0) = 10^4$, $V_{ij}(0) = 0$, $1 \leqslant i, j \leqslant 5$, $i \neq j$, respectively is presented. The numbers indicate the node numbers.

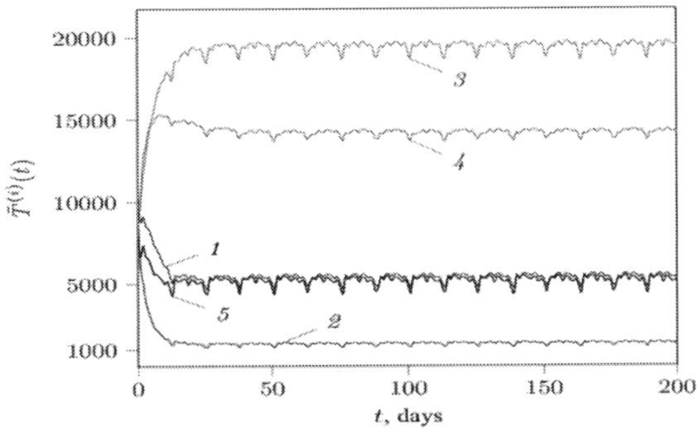

Figure 3. Dynamics of point estimates $T^{(i)}(t)$ without HIV-1 infection at experiment 1.

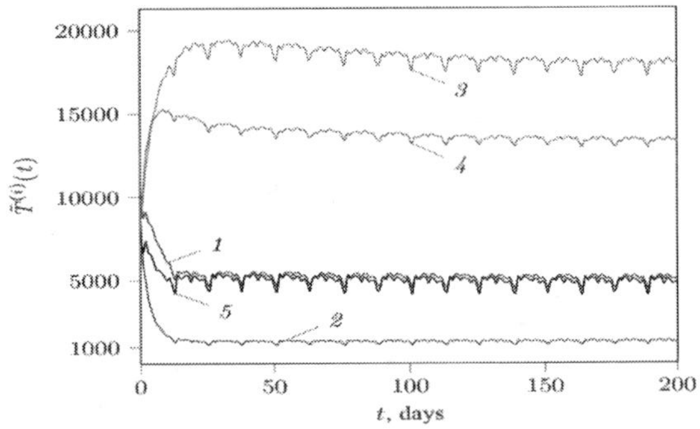

Figure 4. Dynamics of point estimates $T^{(i)}(t)$ with HIV-1 infection at experiment 1.

As can be seen from the figures 3, 4, the number of T-cells decreases slightly in the third and fourth nodes. Let $I(t) = \sum_{i=1}^{5}\sum_{j=1}^{5} \mathbf{E}\, I_{ij}(t)$, $V(t) = \sum_{i=1}^{5}\sum_{j=1}^{5} \mathbf{E}\, V_{ij}(t)$. The figures 5, 6 show the dynamics of the point estimates $I(t)$ and $V(t)$ respectively for $V_{ii}(0) = 10^4$, $V_{ij}(0) = 0$, $1 \leqslant i, j \leqslant 5$, $i \neq j$. In the figures we can see that the numbers of infected cells and virions remains at some positive level for quite a long time.

Experiment 2. The set of parameters from experiment 1 was used, except

$$r_{11}^{(T)} = 3500; \; \gamma_{11}^{(T,V)} = 10^{-7}, \; \gamma_{44}^{(T,V)} = \gamma_{55}^{(T,V)} = 6.5 \cdot 10^{-7};$$
$$\gamma_{33}^{(T,I)} = \gamma_{55}^{(T,I)} = 5 \cdot 10^{-6}; \quad \nu_{55}^{(U)} = 110.$$

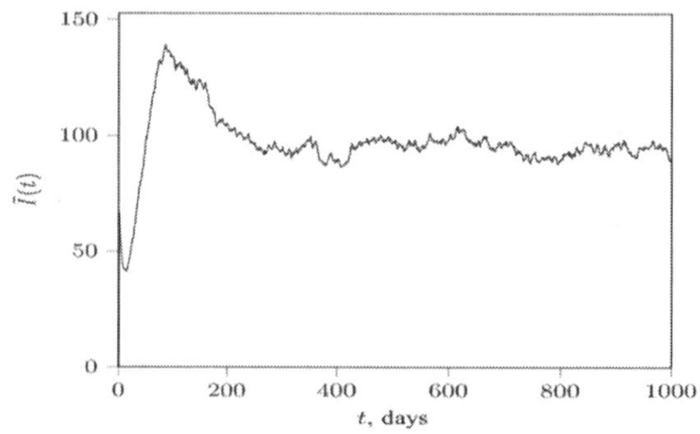

Figure 5. Dynamics of point estimates $I(t)$ at experiment 1.

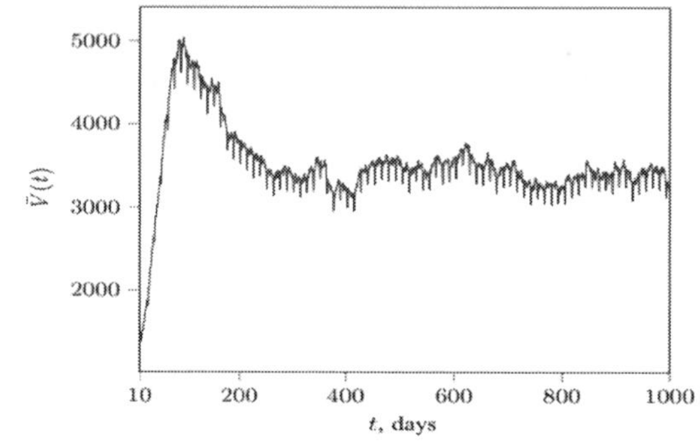

Figure 6. Dynamics of point estimates $V(t)$ at experiment 1.

The quotient basic reproductive numbers are as follows:

$$R_1^{(0)} = 9.85 > 1, \quad R_2^{(0)} = 0.82, \quad R_3^{(0)} = 0.87,$$
$$R_4^{(0)} = 0.89, \quad R_5^{(0)} = 0.73.$$

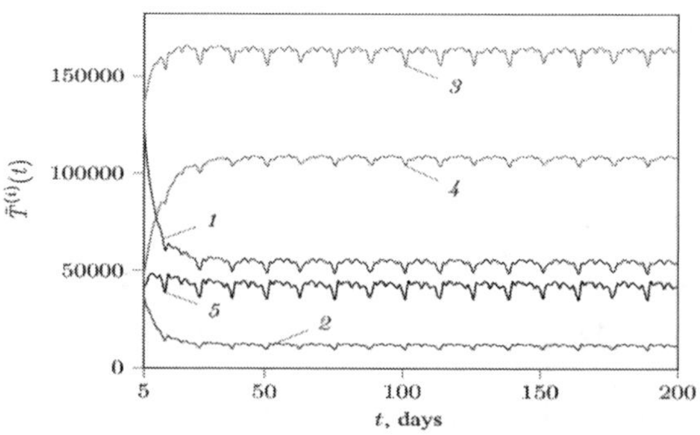

Figure 7. Dynamics of point estimates $T^{(i)}(t)$ without HIV-1 infection at experiment 2.

 https://doi.org/10.1051/itmconf/20203102003

On the figures 7, 8 dynamics of the point estimates $T^{(i)}(t)$ without HIV-1 infection, and with HIV-1 infection ($V_{ii}(0) = 10^4$, $V_{ij}(0) = 0$, $1 \leqslant i, j \leqslant 5$, $i \neq j$), respectively is presented. As can be seen from the figures 7, 8, number of lymphocytes in HIV-infection is significantly

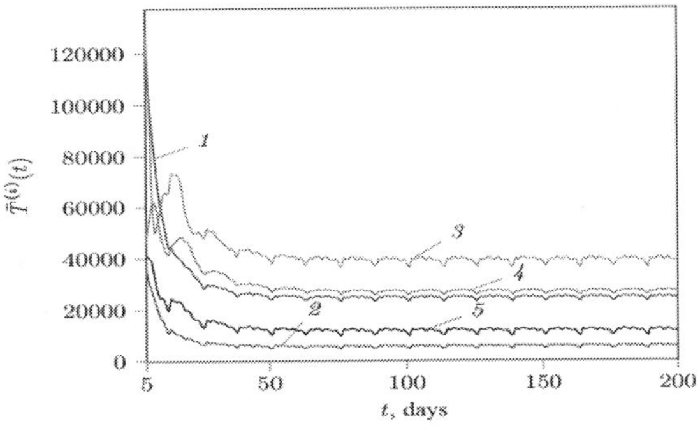

Figure 8. Dynamics of point estimates $T^{(i)}(t)$ with HIV-1 infection at experiment 2.

reduced compared with the absence of HIV-infection. The figures 9, 10 show the dynamics of the point estimates $I(t)$ and $V(t)$ respectively for $V_{ii}(0) = 10^4$, $V_{ij}(0) = 0$, $1 \leqslant i, j \leqslant 5$, $i \neq j$, at experiment 2.

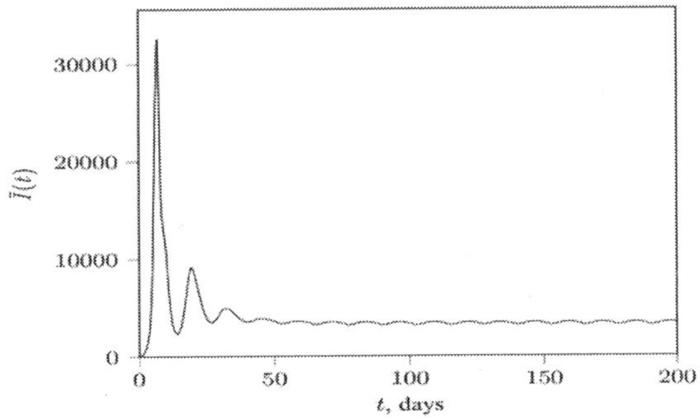

Figure 9. Dynamics of point estimates $I(t)$ at experiment 2.

In the figures 9, 10 we can see that number of infected cells I and virions V after some fluctuations go to some stationary levels. The presence of oscillations is explained by cosines in representation of $\Delta_{ij}^{(I)}(t)$, $\Delta_{ij}^{(V)}(t)$ functions.

5 Conclusions

A continuously-discrete compartmental stochastic model of the dynamics of HIV-1 infection has been developed taking into account the interaction of cells and viral particles in a separate lymph node, the maturation of cells and viral particles, as well as the movements of cells and viral particles through the human lymphatic system. An important feature of the developed continuously-discrete stochastic model is the parameterization of the duration of the movement of particles between the compartments using the specified time functions. Further development of the model may be associated with a complication of the processes

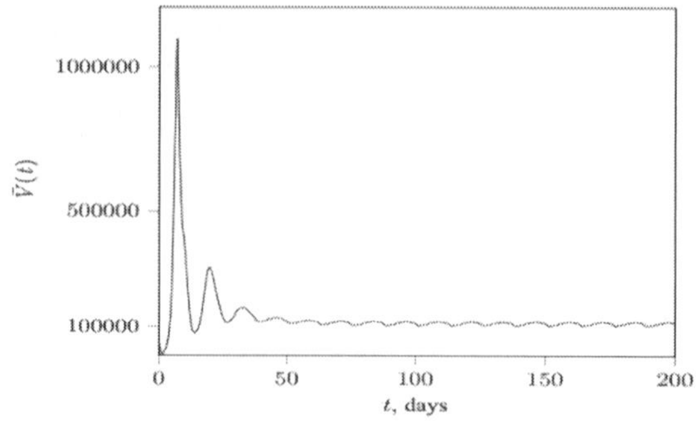

Figure 10. Dynamics of point estimates $V(t)$ at experiment 2.

occurring in the vessels connecting the lymph nodes. In particular, infection of target cells and the production of viral particles by infected cells in the vessels will be considered.

The research was funded by the program of fundamental scientific researches of the Siberian Branch of RAS no. I.1.3.2, project no. 0314-2019-0009.

References

[1] R. Eftimie, J. J. Gillard, and D. A. Cantrell, Bull.Math. Biol. **78**, 2091–2134 (2016).

[2] G. Bocharov, V. Chereshnev, I. Gainova, S. Bazhan, B. Bachmetyev, J. Argilaguet, J. Martinez, and A. Meyerhans, Math. Model. Nat. Phenom. **7 (5)**, 78–104 (2012).

[3] K. A. Pawelek, S. Liu, F. Pahlevani, and L. Rong, Math. Biosci. **235 (1)**, 98–109 (2012).

[4] M. Pitchaimani, C. Monica, J. Appl. Math. Comput. **48**, 293–319 (2015).

[5] D. S. Taltavull, A. Vieiro, and T. Alarcon, J. Math. Biol. **73**, 919–946 (2016).

[6] N. V. Pertsev, B. Yu. Pichugin, and K. K. Loginov, J. Appl. Ind. Math. **13 (1)**, 103–117 (2019).

[7] I. Gyori, J. Eller, Math. Biosc. **53**, 223–247 (1981).

[8] A. S. Mozokhina, S. I. Mukhin, Dif. Eq. **54 (7)**, 938–944 (2018).

[9] S. Nakaoka, I. Shingo, and K. Sato, J. Math. Biol. **72**, 909–938 (2016).

[10] K. K. Loginov, N. V. Pertsev, V. A. Topchii, Math. Biol. Bioinf. **14 (1)**, 188–203 (2019) (in Russian).

[11] R. M. Tretyakova, G. I. Lobov, G. A. Bocharov. Math. Model. Nat. Phenom. **13 (5)**, 44–57 (2018).

[12] R. Tretyakova, R. Savinkov, G. Lobov, G. Bocharov. Computation. **6(1)**, 1–16 (2018).

Modeling the dynamics of population immunity to influenza in Russian cities

Vasiliy N. Leonenko[1,*] and *Daria M.* Danilenko[2,**]

[1]ITMO University, 49 Kronverksky Pr., 197101 St. Petersburg, Russia
[2]Smorodintsev Research Institute of Influenza, 15/17 prof. Popov street, 197376 St. Petersburg, Russia

Abstract. Prediction models which will explicitly include the immunity levels of the population are required to plan effective measures for the containment of seasonal epidemics of influenza. The aim of the current work is to develop an approach to herd immunity dynamics modeling, with the long–term goal of employing it as a part of multicomponent model of influenza incidence dynamics. Based on serological studies performed for 52 Russian cities and 2 virus strains (A(H1N1)pdm09, A(H3N2)) in 11 years period, we propose statistical models which allow to analyze and predict the strain–specific immunity dynamics.

1 Introduction

Epidemic outbreaks of influenza, along with the corresponding increase in the incidence of acute respiratory infections (ARI), cause 3-5 million cases of severe illness and 250-500 thousand deaths annually, according to WHO estimates [1]. Effective planning of measures to contain influenza epidemics requires forecasts of influenza dynamics over the season. In this regard, the world's health authorities started to use prognostic models of ARI and influenza propagation, based on methods of statistical data processing and mathematical modeling, which complement traditional epidemiological surveillance systems. The factors which influence the intensity and timing of seasonal epidemics and thus might be considered in the models are temperature and humidity [2], [3], [4], contact patterns between the individuals [5], [6], biological and antigenic properties of the pathogen [7], [8], and, last but not least, levels of population immunity determined by vaccination coverage and the history of previous influenza infections in the population [9]. In vast majority of works devoted to the modeling of influenza epidemics, the dynamics of the population immunity levels from season to season is either added as an abstract parameter of the model, without its direct assessment from serological studies, — sometimes due to the lack of the corresponding data, — or it is not considered at all. Nevertheless, immunity plays a crucial role in case of repetitive seasonal influenza epidemics (as opposed to the pandemic outbreaks where the immunity could be considered negligible). In particular, due to the growing influence of collective immunity on the dynamics of morbidity, the prognostic model of the spread of influenza in the USSR, which was used in 1970s at the Research Institute of Influenza [10], showed a decline in the accuracy of predictions and due to that reason was put out of service [11]. The calibration of influenza dynamics models to contemporary Russian ARI data [12] confirmed that in case

*e-mail: vnleonenko@yandex.ru
**e-mail: daria.baibus@gmail.com

when ARI incidence is the only type of data used for the calibration procedure, the forecasting accuracy becomes low [13], [14]. Also, the population immunity level might not be accurately assessed by such model calibration when it is treated as a free parameter [15]. This issue badly affects the plausibility of predicted influenza dynamics. In these circumstances, it seems important to incorporate additional sources of data related to immunity levels into the predictive models.

Among the prognostic models that explicitly include the dynamics of collective immunity, we can name the following ones.

- A model with an age structure designed to assess the real number of cases in London during the 2009 influenza pandemic [16].
- A multicomponent model used to analyze the epidemics caused by the A(H1N1)pdm09 influenza strain in the UK in 2009–2011 [17].
- A stochastic model that reproduces the dynamics of influenza in certain regions of England and Wales for 14 years [18].
- A SIRS model taking into account the age structure and differences of individuals in the levels of acquired immunity, calibrated to data of the 2009 pandemic in Hong Kong [19].

Due to the fact that these models were calibrated to data sets gathered at specific locations (and usually within a limited time frame), the conclusions obtained in the mentioned studies cannot be fully generalized to influenza epidemics in other countries, in particular, in the Russian Federation. Also, none of these works demonstrated a comparison of the dynamics of the population immunity in different areas of the country under consideration.

The aim of the presented research is to develop an approach to modeling and forecasting of collective immunity dynamics in urban populations based on the data obtained from biological samples. Our long–term goal is to build multicomponent prognostic models of the spread of influenza epidemics taking into account the levels of the population immunity.

2 Data

The dataset we used to assess the immunity levels is based on serological studies performed for 52 Russian cities and 2 virus strains (A(H1N1)pdm09, A(H3N2)) in 11 years period (2009–2019). Each epidemic season (July to June) two serological samples were analyzed to measure a pre–epidemic (October–November) and a post–epidemic (April–May) immunity levels. The analysis was performed by the local healthcare facilities and verified by the specialists of Research Institute of Influenza, the resulting datasets were stored in .xlsx tables. In a number of results obtained from local healthcare units, incorrect data were identified and excluded. The immunity dynamics to both regarded virus strains in all of the cities is shown in Fig. 1.

Currently, most of the healthcare units do not report age–specific immunity, nor do they distinguish the vaccinated individuals from the non-vaccinated ones when collecting the samples. We assume that there is a need for separate accounting of data for vaccinated and non–vaccinated cohorts, since the antibody responses of these cohorts are determined by distinct reasons and have statistically significant differences. It is supposed to verify this hypothesis as the data on donors with known vaccination status are replenished (currently, the data is only available starting from the 2017–2018 epidemic season).

As it can be seen from the gaps in the graph trajectories, the reporting rate of different cities varies considerably. In some cases the amount of omitted data did not allow assessment of the immunity dynamics, so the corresponding cities were excluded from further analysis. There are 34 cities with the immunity data for both virus strains containing 10 points (i.e. 5

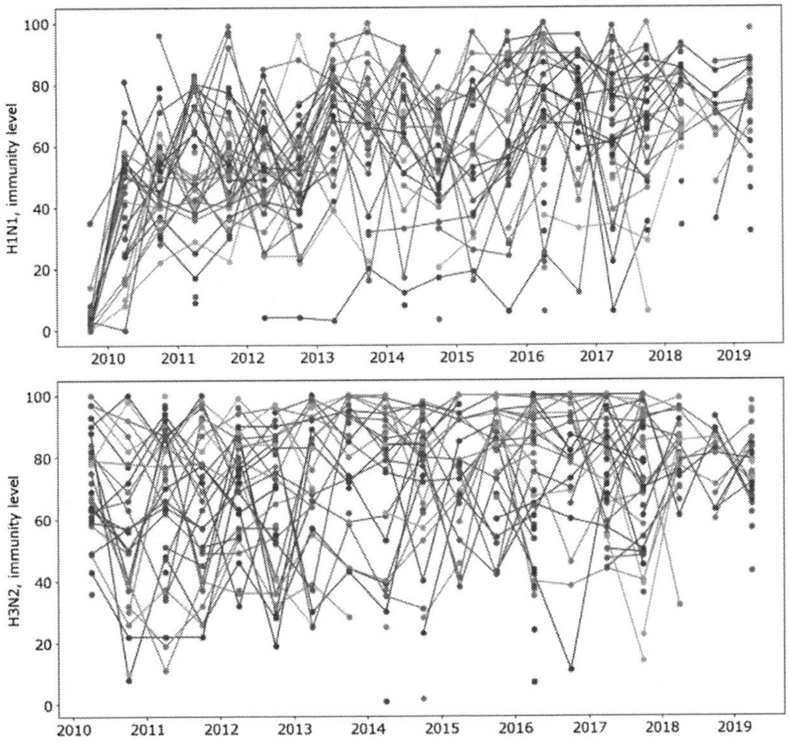

Figure 1. Immunity dynamics to A(H1N1)pdm09 (top) and A(H3N2) (bottom). Each trajectory represents a particular city.

years of measurements) or more. Having added to our consideration less complete data for the cities with the number of data points from 6 to 9, for further analysis we have A(H1N1)pdm09 strain immunity data for 38 cities and A(H3N2) strain immunity data for 39 cities.

3 Analyzing immunity dynamics

Based on the selected data, we analyzed the dynamics of collective immunity over the years and highlighted significant factors affecting the levels of population immunity. It was revealed that, depending on the dominant subtype of influenza viruses in the epidemic, fluctuations in the level of collective immunity are observed in the pre– and post–epidemic periods. In particular, in the post–epidemic period, there is an increase in the number of seropositive donors to the subtype that is most intensively circulating during the epidemic. Also, the preliminary analysis results in a hypothesis that the year of the onset of circulation of the strain in question has a great influence on the change in population immunity levels. Indeed, there is no city to demonstrate immunity levels close to 100% to A(H1N1)pdm09 in the first years of observation (see Fig. 1, top). The immunity levels to A(H3N2) are much higher and almost do not have a tendency to grow (see Fig. 1, bottom).

Next was the stage of constructing a model of the immunity dynamics. A review of mathematical methods suitable for modeling the dynamics of immunity was conducted. Three approaches were considered as candidates: a mathematical model based on discrete difference equations, regression analysis, and a recurrent neural network. Guided by the results of the review, as well as the results of data analysis (in particular, by the estimated samples size and by available knowledge from the field domain), we chose an approach to short–term forecasting of levels of population immunity based on the regression analysis. A software module was created for predicting the dynamics of population immunity using the retrospective data. For this purpose, we have chosen Python programming language, because it is freely available and allows easy batch processing of data, unlike many GUI–oriented statistical software tools. The module was initially tested on average data on the levels of population immunity in the Russian Federation to strains of A (H1N1)pdm09 and A (H3N2) for 2009–2017 (two measurements each year) which were available from the open sources [9], [20] (see Fig. 2). The prediction accuracy assessment was carried out by predicting the points excluded from the test set (see Fig. 2, light–blue dots). The number of excluded points varied. The module was searching for the best regression function, selecting it from the several predefined functions (based on combinations of logarithmic functions, trigonometrical functions and polynomial terms). The accuracy comparison was made based on the two output characteristics: values of the coefficient of determination R^2 demonstrated for all the points, including the predicted ones, and the values of mean squared error (MSE) for the predicted points. The first parameter is aimed at characterizing the ability of the selected model to describe the whole dataset, whereas the second one demonstrates the predictive ability of the model. The best result corresponds to the maximal R^2 and minimal MSE.

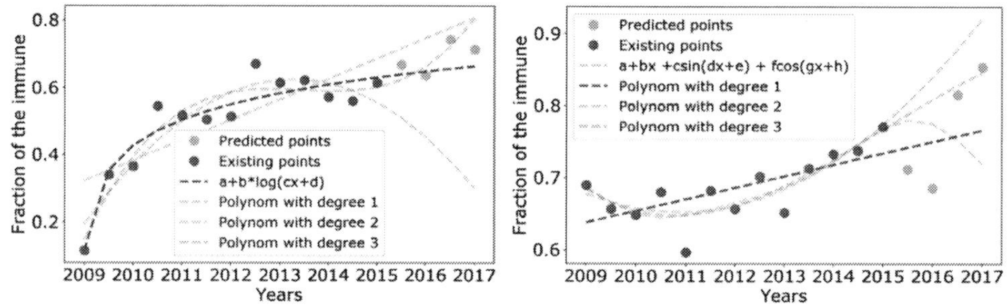

Figure 2. The prediction of the average immunity levels for the population of Russia (four last points in time series) using different regression models.

The obtained selection of functions to describe the dynamics of average all-Russian immunity levels to A(H1N1)pdm09 and A(H3N2) over time was verified against the data for particular Russian cities, described in the section 2. The examples of the results are shown in Fig. 3 and Table 1.

	log	deg.1	deg.2	deg.3
R^2	**0.63**	0.35	0.48	-0,73
MSE	**0.04**	0.11	0.06	0.56

	sin/cos	deg.1	deg.2	deg.3
R^2	**0.35**	0.27	0.34	0.14
MSE	0.06	0.02	**0.001**	0.11

Table 1. Fitting and predictive accuracy of different models for the immunity to A(H1N1)pdm09 in Kursk (left table) and A(H3N2) in Vologda (right table). The optimal values are shown in bold.

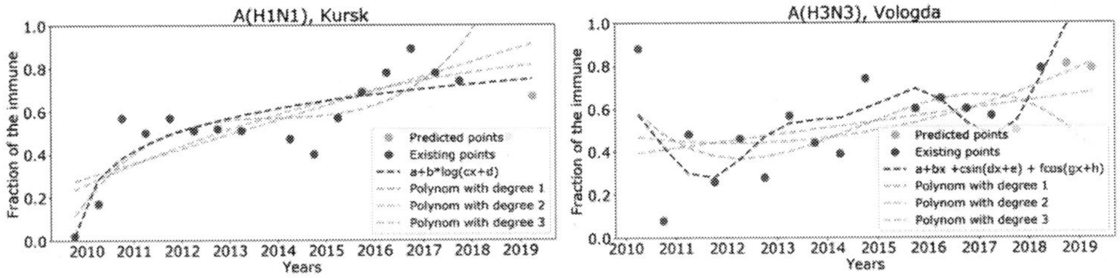

Figure 3. The prediction of the immunity levels for the population of Russian cities (two last points in time series) using different regression models, with data for Kursk and Vologda as an example.

4 Results and discussion

Based on the output, we conclude that the logarithmic function demonstrated better accuracy for predicting immunity to A (H1N1) pdm09 among the tested regression models, being the best by R^2 value for 17 cities and by MSE value for 13 cities out of 38. The reason is in the fast growth of immunity levels to this strain due to its seasonal circulation after the pandemic outbreak in 2009. The selection of the best model is more difficult in case of A (H3N2). The levels of immunity to A (H3N2) demonstrate fluctuations around the linear trend (close to the horizontal line in some cities). These fluctuations are better captured by the periodic model, – best fit by R^2 for 17 cities out of 38, – but due to their irregular nature, the linear trend function has much more predictive power (the lowest MSE for 22 cities). The achieved results are consistent with the previously formulated hypothesis that the dynamics of immunity depends on the year the strain began to circulate: strain A (H1N1) pdm09 began to circulate in 2009, as a result of which the level of population immunity to it increases in time, while population immunity to strains A (H3N2) should be more stable due to the circulation of viruses of this subtype from the late 60s. It is known that epidemiologists observe that in the absence of virus circulation in a population, immunity to it begins to decline, and its drop to a critical level leads to an epidemic of influenza caused by this strain. For example, the work [21] provides an estimate of this level at 0.5, i.e., 50% of the population should not have immunity to the circulating strain. The indicated effect is not observed on the data set available to the authors. In the future, it will be interesting to consider a sample with laboratory tests of immunity for a longer period of time and to verify the mentioned observation quantitatively.

The obtained statistical models will be used to perform data imputation to assess the immunity levels which were not reported by the healthcare units. Also the mentioned models will be incorporated into the models of influenza dynamics to enhance their descriptive and predictive force. For this case, it should be considered that the data on immunity levels we regarded in this study are related solely to biological immunity, whereas the factor of the indirect protection due to peculiarities of the contact patterns of the population cannot be measured in such a way. This fact might result in underestimating of the real immunity level when using the statistical data in the models. To take into the account the immunity connected with indirect protection, spatially explicit models could be used which consider the contact structure of the population [22].

5 Acknowledgments

The work was supported by Russian Foundation for Basic Research [grant number 18-37-00428].

References

[1] WHO, *Influenza (seasonal). Fact sheet No. 211, March 2014.*, [online], http://www.who.int/mediacentre/factsheets/fs211/en/

[2] A. Romanyukha, T. Sannikova, I. Drynov, Herald of the Russian Academy of Sciences **81**, 31 (2011)

[3] S.P. van Noort, R. Águas, S. Ballesteros, M.G.M. Gomes, Journal of Theoretical Biology **298**, 131 (2012)

[4] V.N. Leonenko, S.V. Ivanov, Y.K. Novoselova, Procedia Computer Science **80**, 2402 (2016)

[5] L.M. Bettencourt, Science **340**, 1438 (2013)

[6] S. Kumar, K. Piper, D.D. Galloway, J.L. Hadler, J.J. Grefenstette, BMC public health **15**, 947 (2015)

[7] J. Tamerius, M.I. Nelson, S.Z. Zhou, C. Viboud, M.A. Miller, W.J. Alonso, Environmental health perspectives **119**, 439 (2011)

[8] R. Yaari, G. Katriel, A. Huppert, J. Axelsen, L. Stone, Journal of The Royal Society Interface **10**, 20130298 (2013)

[9] O. Konshina, A. Sominina, E. Smorodintseva, K. Stolyarov, I. Nikonorov, Russian Journal of Infection and Immunity **7**, 27 (2017), in Russian.

[10] *Research Institute of Influenza website*, [online], http://influenza.spb.ru/en/

[11] Y. Ivannikov, P. Ogarkov, Journal of Infectology **4**, 101 (2012), In Russian.

[12] V.N. Leonenko, S.V. Ivanov, Russian Journal of Numerical Analysis and Mathematical Modelling **31**, 267 (2016)

[13] V.N. Leonenko, Y.K. Novoselova, K.M. Ong, Procedia Computer Science **101**, 282 (2016)

[14] V.N. Leonenko, S.V. Ivanov, Mathematical Biosciences and Engineering **15**, 209 (2018)

[15] V. Leonenko, G. Bobashev, Epidemics **29**, 100358 (December 2019)

[16] P.J. Birrell, G. Ketsetzis, N.J. Gay, B.S. Cooper, A.M. Presanis, R.J. Harris, A. Charlett, X.S. Zhang, P.J. White, R.G. Pebody et al., Proceedings of the National Academy of Sciences **108**, 18238 (2011)

[17] I. Dorigatti, S. Cauchemez, N.M. Ferguson, Proceedings of the National Academy of Sciences **110**, 13422 (2013)

[18] M. Baguelin, S. Flasche, A. Camacho, N. Demiris, E. Miller, W.J. Edmunds, PLoS medicine **10**, e1001527 (2013)

[19] H.Y. Yuan, M. Baguelin, K.O. Kwok, N. Arinaminpathy, E. van Leeuwen, S. Riley, Epidemics **20**, 84 (2017)

[20] M. of health of Russian Federation, *Kharakteristika epidemii grippa sezona 2017-2018 gg. v rossiyskoy federatsii. [characteristics of the influenza epidemic season 2017-2018 in russian federation.]* (2018), in Russian.

[21] L. Semenova, *Epidemiologicheskiye osobennosti sovremennogo grippa i sovershenstvovaniye taktiki vaktsinatsii naseleniya [epidemiological features of modern influenza and improving the vaccination tactics of the population]. the dissertation for the degree of candidate of medical sciences.* (2012), in Russian.

[22] V. Leonenko, A. Lobachev, G. Bobashev, *Spatial Modeling of Influenza Outbreaks in Saint Petersburg Using Synthetic Populations*, in *International Conference on Computational Science* (Springer, 2019), pp. 492–505

Analyzing the spatial distribution of individuals predisposed to arterial hypertension in Saint Petersburg using synthetic populations

Vasiliy N. Leonenko[1,*] and *Sergey V.* Kovalchuk[1]

[1]ITMO University, 49 Kronverksky Pr., St. Petersburg, 197101, Russia

Abstract. Arterial hypertension (AH) is one of the most common cardiovascular diseases, and it can lead to serious complications. To optimize the delivery of patients exposed to AH to medical institutions and thus to curtail mortality in Russian cities caused by the consequences of hypertension, it is necessary to estimate the number of potential patients, along with their spatial distribution. This paper presents a method which uses synthetic population data to assess the spatial distribution of individuals potentially prone to arterial hypertension. The risk of arterial hypertension of an individual is calculated based on its demographic characteristics (age and gender). Using Saint Petersburg as a case study, we demonstrate that the mentioned approach makes it possible to perform predictions of AH cases distribution in absence of real data on hypertension status of the individuals. The results of the study will be used to assess the input flows of patients to healthcare facilities and optimize their workflow.

1 Introduction

Arterial hypertension (AH), a major cause of premature deaths worldwide, is a medical condition associated with elevated blood pressure. For 2019, hypertension affects around 1.13 billion people. The aim of reducing the prevalence of hypertension by 25% by 2025 is one of the global targets for noncommunicable diseases [1].

To prevent mortality of individuals from the consequences of hypertension in Russian cities, it is necessary to estimate the number and the spatial distribution of city dwellers exposed to hypertension. Such information will help to optimize the delivery of patients to medical institutions and their treatment [2].

This paper presents an algorithm for modeling the spatial distribution of individuals exposed to arterial hypertension. The proposed algorithm assesses the distribution of individuals prone to hypertension based on the synthetic populations technique coupled with the statistical model of the incidence of AH [3] and Monte Carlo methods. We demonstrate the output produced by the algorithm using the population of Saint Petersburg as a case study.

2 Synthetic population

A "synthetic population" is a synthesized, spatially explicit human agent database (essentially, a simulated census) representing the population of a city, region or country. By its

*e-mail: vnleonenko@yandex.ru

cumulative characteristics, this database is equivalent to the real population, but its records does not correspond to real people. Statistical and mechanistic models built on top of the synthetic populations helped tackle a variety of research problems, including those connected with public health. One of the standards of synthetic populations is proposed and developed by RTI International [4]. It was used by various research groups to create populations for 50 US states, along with another regions and countries. The authors have experience of working with the synthetic populations from RTI, using it for statistical analysis of opioid–related overdoses in Cincinnatti [5] and for the investigation of the dynamics of influenza in Russian cities [6].

According to the standard of RTI International, a synthetic population consists of several txt-files, each of them containing a table with every row being a single record corresponding to some entity – an individual, a household, a workplace, a school, etc. The principal data is stored in four files: `people.txt` (each record contains id, age, gender, household id, workplace id, school id), `households.txt` (contains id and coordinates), `workplaces.txt` (contains id, coordinates and capacity of the workplaces), `schools.txt` (contains id, coordinates, capacity). Using the RTI synthetic population protocol as a reference, we generated the corresponding files for the population of Saint Petersburg. Since the data available for Saint Petersburg was not complete, we altered or outright omitted some of the methods used in the original procedure. That resulted in a synthetic data, which might not exactly match the real population. We regard these data as an initial approximation and we plan to make our synthetic population more accurate employing additional sources of information.

The principal data source for our synthetic population is 2010 data from "Edinaya sistema ucheta naseleniya Sankt Peterburga" ("Unified population accounting system of Saint Petersburg") [7]. The data is represented in a form of Excel spreadsheets containing records with house addresses and the corresponding number of dwellers of certain age and gender. To match the household addresses with the geographical coordinates and assess the plausibility of the obtained geographical data, a computational algorithm was developed and implemented using Python programming language. The named algorithm employed a geocoder to detect locations of the addresses and filtered out errors in the results with multiple addresses matching only one pair of coordinates. In the same fashion the coordinates of schools were assigned, based on the school list from the official web–site of the Government of Saint Petersburg [8].

The distribution of working places for adults and their coordinates were derived from the data obtained with the help of Yandex.Auditorii API [9]. Initially the data was available in a form of a file in GeoJSON format (it is an open standard format designed for representing simple geographical features, along with their non-spatial attributes). The file consisted of relative workplace size assessments for each of the cells in a hexagonal grid. This data was normalized using the official cumulative employment numbers [10]. Synthetic workplace records were created by assigning the calculated number of employees in each hexagonal cell to imaginary geographical location coinciding with the center of this cell.

We assumed that young people aged 7 to 17 attend schools, and the adults of working age (18 to 55 for females and 18 to 60 for males) might be working. Iterating through the list of records for the city dwellers generated earlier, we were assigning each person to a closest school or working place, until they are filled to capacity or there is no more people to be assigned. More details on the population used and the algorithms for its generation employed could be found in [6].

3 Assessing AH risk and individual AH status

When the synthetic population is created, we assess the health conditions of individuals associated with arterial hypertension. There are two types of corresponding data that we generate and add to the individual records of the synthetic population:

- The AH risk (the probability of having arterial hypertension). Based on [3], we assumed that the mentioned probability depends on age and gender of an individual. The corresponding cumulative distribution function is shown in Fig. 1. The example of individuals distribution on the map with their AH risk divided into three categories (low, average and high) is shown in Fig. 2.
- The actual AH status (positive or negative). The corresponding value (0 or 1) is generated by the Monte Carlo algorithm according to the AH risk calculated in the previous step. The AH status might be used in simulation models which include demographic processes and population-wide simulation of the onset and development of AH.

At the moment, none of the parameters of the synthetic population were used for the hypertension status assessment aside from age and gender, but we consider adding more factors to the statistical model, where the workplace distribution information might also be utilized. For example, adding workplace type into the synthetic population data will allow us to assume the amount of daily physical activity of individuals working there, which is one of the factors correlated (inversely) with the AH probability [1]. Also, the geographical locations of the individuals' workplaces may be used in simulation models to calculate the distribution of cases of acute conditions ocurring in working hours and thus to facilitate balancing of the incoming flows to healthcare facilities.

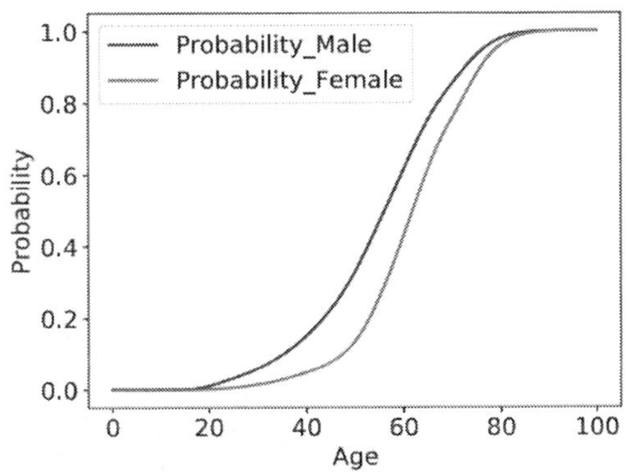

Figure 1. The cumulative distribution function used to define the AH status of an individual, based on data from [3].

4 Results

To visualize the spatial distribution of the individuals prone to AH (shortly, AH+ individuals), we convert the coordinates of an individuals' household locations from degrees to meters using Mercator projection. After that we form a grid with cell size 250m x 250m, defining its bounds with maximum and minimum coordinates of the regarded synthetic individuals. Then we calculate the overall number of dwellers and the number of individuals exposed to AH for each cell of the grid. The algorithm was implemented as a collection of scripts written

Figure 2. The distribution of individuals of different risk categories to arterial hypertension

in Python programming language with the libraries `numpy`, `matplotlib`, and `pandas`. The output of the algorithm is a txt-file with the coordinates of the cells and the cell statistics (overall number of individuals, number of AH+ individuals). As a final step, we use QGIS open software to draw cells on the map of Saint Petersburg (OpenStreetMap background layer). The result is demonstrated in Fig. 3.

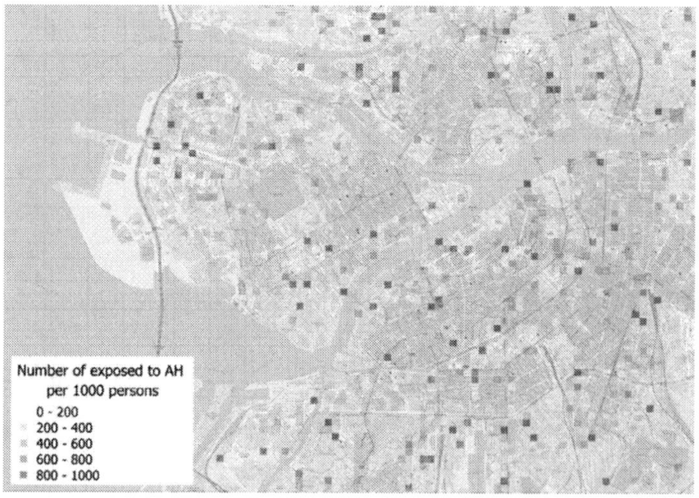

Figure 3. Density of the individuals exposed to arterial hypertension in the cells with size 250m x 250m

As it is seen from the map, the discrete representation of the density of individuals with arterial hypertension sometimes leads to sharp borders between the cells, which is not plausible. Also, the selection of the cell size strongly affects the visualization outcome. To resolve these issues, we used an interpolation function in QGIS to smoothen the transition between the regions with different densities. In Fig. 4, we demonstrated a part of the city map corresponding to Vasilyevsky island, one of the central districts of Saint Petersburg, with the interpolated data on the density of AH+ individuals.

The proportion of synthesized AH+ individuals in the population obtained with the help of our algorithm is not uniformly distributed in space. Particularly, it can be seen in Fig. 4 that there are certain spots of high density of the people with AH in the south and north–west, and there is a spot of low density of AH in the south–west of the island. Since the statistical model utilized to assess the AH risk considers only age and gender as AH development factors, the heterogeneity of AH+ clusters basically reflects the heterogeneity of demographic parameters distribution of the population. To confirm or deny the hypothesis that those factors are essential for AH assessment and that our method reproduces AH+ distribution with satisfactory accuracy, additional spatially explicit datasets should be used. We plan to use the hospital data on emergency calls in Saint Petersburg in 2015 which were related to acute coronary syndrome (ACS) [11]. Since ACS is one of the most widespread consequences of AH [1], our idea is to compare different neighborhoods by the value of the ratio of total number of ACS cases to the assessed number of AH+ individuals. The same technique was earlier developed and applied by the authors to the ratio between the number of calls to emergency services related to opioid overdoses and the assessed number of opioid drug users in Cincinnatti [5].

Figure 4. Assessment of the density of AH+ individuals in Vasilyevsky island, St Petersburg, using the interpolated data

5 Discussion

In this study we presented an approach based on the synthetic population concept that makes it possible to perform analysis of the spatial distribution of individuals predisposed to arterial hypertension. The algorithms for assessing and visualizing the density of city dwellers prone to AH are implemented using Python and QGIS open software. It is worth noting that the approach could be used to analyze the distribution of any chronic condition if the corresponding population data is available. The next planned improvement is to implement hotspot analysis based on Getis–Ord Gi* statistics and to mark statistically significant anomalies on the map (unusually high and unusually low density of AH cases compared to the city average). Also, a separate algorithm will be developed to assess the prospected incoming flow of patients with AH to a hospital in a fixed map location.

There is a number of drawbacks of the study which should also be addressed, namely:

- At the moment we use the synthetic population based on the 2010 census data. This population does not take into account the newly constructed housing and the increased number of dwellers in Saint Petersburg. We plan to employ the 2018 data to bring the population up to date.
- The current visualization approach makes it somewhat difficult to distinguish the areas with low density of AH cases and scarcely populated areas (for instance, industrial zones). The way to overcome this issue might be in regarding a special indicator instead of AH density.
- The statistical model could be modified to include the dependency of AH probability on additional parameters, apart from age and gender, for instance, alcohol and tobacco consumption and body mass index of a person [1].

We assume that these modifications will further enhance the accuracy of the algorithm and will make it a useful tool for the healthcare specialists.

6 Acknowledgements

This research is financially supported by The Russian Science Foundation, Agreement #19-11-00326. The authors thank Dr. Anna Semakova for providing the data for the cumulative distribution function which was used to assign the arterial hypertension status.

References

[1] WHO, *Hypertension. Fact sheet*, [online], https://www.who.int/news-room/fact-sheets/detail/hypertension

[2] S.V. Kovalchuk, A.A. Funkner, O.G. Metsker, A.N. Yakovlev, Journal of Biomedical Informatics **82**, 128 (2018)

[3] A. Semakova, N. Zvartau, Procedia Computer Science **136**, 433 (2018)

[4] W.D. Wheaton, J.C. Cajka, B.M. Chasteen, D.K. Wagener, P.C. Cooley, L. Ganapathi, D.J. Roberts, J.L. Allpress, Methods report (RTI Press) **2009**, 905 (2009)

[5] S. Bates, V. Leonenko, J. Rineer, G. Bobashev, Computational and Mathematical Organization Theory **25**, 36 (2019)

[6] V. Leonenko, A. Lobachev, G. Bobashev, *Spatial Modeling of Influenza Outbreaks in Saint Petersburg Using Synthetic Populations*, in *International Conference on Computational Science* (Springer, 2019), pp. 492–505

[7] Government of Saint Petersburg, *Labor and employment committee. information on economical and social progress*, [online], http://rspb.ru/analiticheskaya-informaciya/razvitie-ekonomiki-i-socialnoj-sfery-sankt-peterburga/ (In Russian)

[8] Government of Saint Petersburg, *Official web-site*, [online], https://www.gov.spb.ru/

[9] Yandex, *Auditorii*, [online], https://audience.yandex.ru/

[10] Government of Saint Petersburg, *Edinaya sistema ucheta naseleniya Sankt Peterburga (Unified population accounting system of Saint Petersburg)*, [online], https://reestr-gis.spb.ru (In Russian)

[11] S.V. Kovalchuk, M.A. Moskalenko, A.N. Yakovlev, *Towards model-based policy elaboration on city scale using game theory: application to ambulance dispatching*, in *International Conference on Computational Science* (Springer, 2018), pp. 404–417

Mathematical modelling on the control of HIV/AIDS with campaign on vaccination and therapy

David Omale[1,*]

[1]Mathematical Sciences Department, Faculty of Natural Sciences, Kogi State University, Anyigba, Kogi State, Nigeria.

Abstract. HIV/AIDS is a global pandemic ravaging nations with negative adverse effect creating high level of mortality rate and increase the number of orphans worldwide. The control of this pandemic disease like any other diseases becomes a global fight, thus we use mathematical epidemic model with effective public health campaign on the need for vaccination and effective therapeutic dose as a measure to it spread. A Susceptible, Vaccinated, Infected, Treatment and AIDS ($SVITA$) compartmental model is developed in the study. We conducted a qualitative analysis on the models by carrying out the local and global stability analysis on the disease free and endemic equilibrium point. The basic reproduction number R_0 is found using the next generation matrix approach. The condition $R_0 < 1$ indicates that with effective public health campaign on the need for vaccination and therapeutic dose, the HIV/AIDS burden will be drastically reduced together with the transmission rate of HIV/AIDS within the population.

1 Introduction

Human immunodeficiency virus infection and acquired immune deficiency syndrome HIV/AIDS is a spectrum of conditions caused by infection with the human immunodeficiency virus HIV [1]. Since the discovery of HIV/AIDS the disease had caused a lot of havoc in the world, most especially in Africa and some less develop countries. HIV is transmitted by three main routes: Sexual contacts, exposure to infected body fluids or tissues and from mother to child during pregnancy delivery or breast feeding(Vertical transmission [2]). HIV/AIDS can be prevented through safe sex, needle exchange programmes, treating those who are infected and male circumcision [3]. HIV is believed to have originated in West Central Africa during the late 19th or 20th century [5]. AIDS was first recognised by the United State Centre for Disease Control and Prevention CDC in 1981 Gallo (2006). Since its inception it has caused an estimated 35 million death worldwide and it is consider a pandemic disease [4].

Vaccine aims to induce an immune response against HIV in which antibodies against HIV are administered [6]. There are no licensed HIV/AIDS vaccine on the market currently, but there is hope that the vaccine will be possible soon, because one HIV vaccine regime RV – 144 has been shown to prevent HIV in some individual in Thailand.

Highly Active Antiretroviral Therapy (HAART) decreases the patient's total burden of HIV, burst the immune system and prevent any onset of opportunistic infections that often lead to death. HAART also prevent the transmission of HIV/AIDS (Eisinger et al (2019)).

*e-mail: davidubahi@gmail.com

Public health education to the public had become the main strategies to curb the spread of this pandemic disease. Campaign on the need for vaccination on the part of those who are currently tested negative to HIV virus and therapeutic dose to those tested positive will go a long way to halt the progressive rate of HIV/AIDS infection globally. Owen et al (1998) used a single risk group model of the epidemic among gay men in San Francisco to examine the effects of prophylactic and therapeutic vaccines [9]. Blower et al (2002) uses a detailed model of vaccination response to investigate the effects of a number of vaccine characteristics on impact [10]. Gray et al (2003) analyzed the effect of ART and vaccination on HIV transmission using data from a cohort study in Rokai, Uganda. Mudandavire (2009) proposed a system of discreate time delay differential equation in modelling effects of public health education campaign on HIV/AIDS transmission with no other intervention available [11]. This work differs from all others in the sense that we consider the effect of public health campaign on vaccination for the active age group who are not yet positive and education on the HAART consistency on those infectious group within the population, using a deterministic differential models Ratera et al (2012), study the effect of screening and treatment on transmission of HIV/AIDS infection in a population, which is quite different from the present approach, they show that screening the unaware HIV infective and treatment of screened HIV infective have the effect of reducing the rate of transmission of HIV disease. Gumel et al (2002) present a mathematical model to design and monitor the interaction between HIV, CD4+ T-Cells, CD8+ T-Cell productively infected and latently infected CD4+ T-Cell and to evaluate therapeutic strategies during the early state of HIV/AIDS infection within the cell [13].

2 Model description and analysis

The model is based on the following assumptions:

1. The population under consideration is homogeneous
2. In the population, there is no vertical transmission, the population consist of active sex adult.
3. And that those in the active aids group doesn't transmit the virus because they may be too ill to be involved in active sex.
4. That vaccine is effective

The model is a deterministic epidemic models with the general population N consist of $N = S + V + I + T + A$ classes, where S is the susceptible, V the vaccination class, I represent the infectious class, T is the treatment class of the infectious and A represents the active AIDS class. The population is being recruited by the parameter Λ into the susceptible population, and through aggressive public education, the susceptible population go for vaccination at the $(1 + w_1)$. The susceptible will be infected with the force of infection $\psi = \frac{(C_1\beta_1+C_2\beta_2)S}{N}$ into the infectious and treatment class respectively, γ represent the rate that infected individual developed a full blown AIDS and $(1 - \gamma)$ represent the rate that some proportion of the infected class under treatment with the public health campaigns w_2. σ represent treatment failure which may caused those in the infected class to developed full blown AIDS. The AIDS class received treatment at the rate θ. All the classes is being reduced by natural death rate denoted by μ and α is the diseases induced death rate. $(1 + w_2)$ represents public health campaign for the infected class to undergo treatment and $(1 + w_1)$ represents the public health campaign on vaccination for the susceptible class and ϕ is the wanning rate of vaccination.

Consider the model flow diagram below

https://doi.org/10.1051/itmconf/20203103003

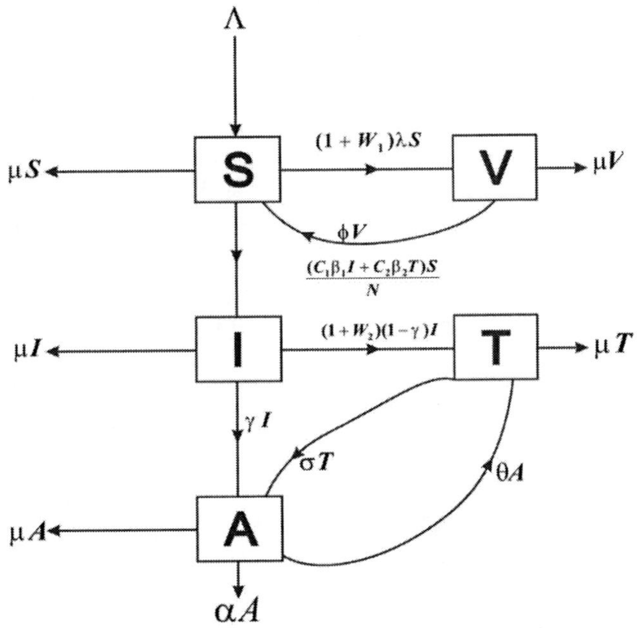

Figure 1. Model Diagram

The mathematical model arising from the above schematic diagram is:

$$\begin{aligned}
\frac{dS}{dt} &= \Lambda - (1+w_1)\lambda S - \left(\frac{C_1B_1I + C_2B_2T}{N}\right)S + \phi V - \mu S.\\
\frac{dV}{dt} &= (1+w_1)\lambda S - \phi V - \mu V.\\
\frac{dI}{dt} &= \left(\frac{C_1B_1I + C_2B_2T}{N}\right)S - (1+w_2)(1-\gamma)I - \gamma I - \mu I.\\
\frac{dT}{dt} &= (1+w_2)(1-\gamma)I + \theta A - \sigma T - \mu T.\\
\frac{dA}{dt} &= \gamma I + \sigma T - \theta A - \alpha A - \mu A.
\end{aligned} \tag{1}$$

Therefore, the disease free equilibrium point is given as

$$\varepsilon_0 = \left(\frac{\Lambda}{[(1+w_1)\lambda + \mu - \frac{\phi}{(\phi+\mu)}(1+w_1)\lambda]}, \frac{(1+w_1)\lambda\Lambda}{[\phi+\mu][(1+w_1)\lambda + \mu - \frac{\phi}{(\phi+\mu)}(1-w_1)]}, 0, 0, 0 \right). \tag{2}$$

Based on the mathematical model above, the endemic equilibrium is attained when $S = S^*, V = V^*, I = I^*, T = T^*, A = A^*$

The endemic state is therefore represented by:

$$\varepsilon_1 = \left(\frac{\Lambda}{[(1+w_1)\lambda + \psi + \mu - \psi(1+w_1)\lambda]}, \frac{(1+w_1)\lambda\Lambda}{[\psi+\mu][(1+w_1)\lambda + \psi + \mu - \psi(1+w_1)\lambda]}, \right.$$
$$\frac{\Lambda\psi}{[(1+w_1)\lambda + \psi + \mu - \phi(1-w_1)\lambda][(1+w_2)(1-\gamma)+\gamma+\mu]},$$
$$\frac{[\gamma(\sigma+\mu)+\sigma(1+w_2)(1-\gamma)]I^*}{(\theta+\alpha+\mu)(\sigma+\mu)-\mu\theta},$$
$$\left. \frac{[(1+w_2)(1-\gamma)[(\theta+\alpha+\mu)(\sigma+\mu)-\sigma\theta]+\theta(\gamma(\sigma+\mu)+\sigma(1+w_2)(1-\lambda))I^*]}{[(\theta+\alpha+\mu)(\sigma+\mu)-\sigma\theta][\sigma+\theta]} \right). \quad (3)$$

Consequently, the basic reproduction number was obtained using the next generation method as;

$$R_0 = \frac{C_1\beta_1}{[(1+w_2)(1-\gamma)+\gamma+\mu]} + \frac{[(1+w_2)(1-\gamma)(\theta+\alpha+\mu)+\theta\gamma]C_2\beta_2\mu}{[(1+w_2)(1-\gamma)+\gamma+\mu][(\sigma+\mu)(\theta+\alpha+\mu)+\theta\sigma][(1+w_1)\lambda+\mu-\phi(1+w_1)\lambda]}. \quad (4)$$

3 Numerical Simulations

In this section, Matlab software was used to simulate the models. In order to show the analytical results of the work, numerical simulations of the model system are carried out using a set of values.

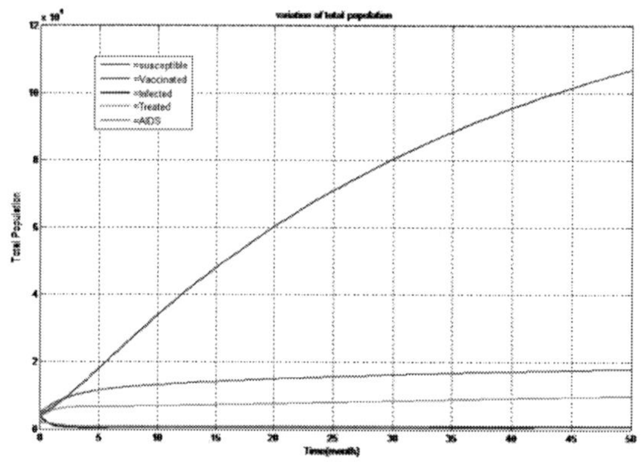

Figure 2. Variation of total population with time

Figure (2) above shows the variation of the proportion of total population in all classes with time using the followng parameters; $\Lambda = 5000$, $w_1 = 0.01$, $w_2 = 0.25$, $\lambda = 0.35$, $\gamma = 0.40$, $(1-\gamma) = 0.60$, $\sigma = 0.061$, $\phi = 0.41$, $\psi = 0.025$, $\beta_1 = 0.020$, $\beta_2 = 0.010$, $C_1 = 0.002$, $C_2 = 0.025$, $0.015 \leq \mu \leq 0.025$ and $0.30 \leq \alpha \leq 0.45$
With serious campaign for vaccination and therapy as seen above, the proportion of susceptible increases, there is a significant decrease in the AIDS and infected classes and increase in vaccination class. The treatment class show an increase and finally dropped.

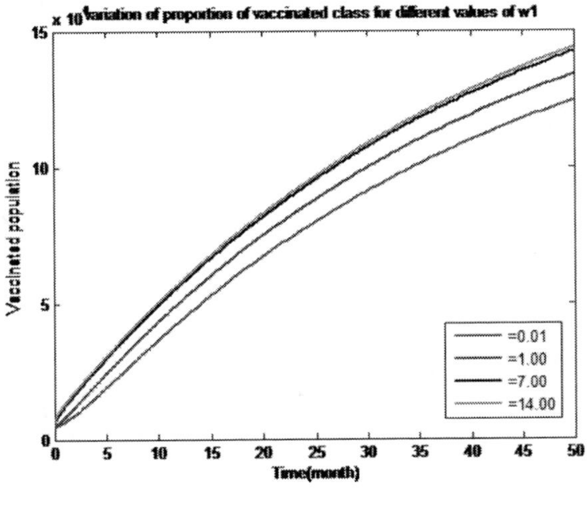

Figure 3. Variation of proportion of vaccinated class for different values of w_1

Figure (3) above shows the variation of proportion of vaccination class for different value of public health campaign w_1 with parameter values ($w_1 = 0.01, 1.00, 7.00, 14.00$). this shows that increase in the public health campaign will lead to increase in vaccination and many people will be covered, when campaign is vigorously increase, it will reduce stigmatisation of people living with the virus.

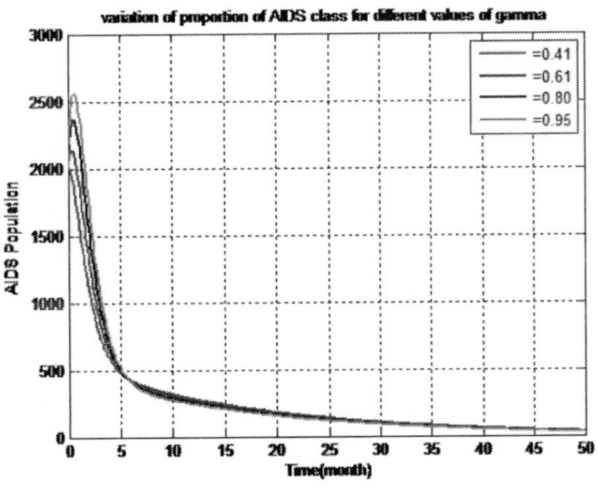

Figure 4. Variation of proportion of AIDS class for different values of γ

Figure (4) above shows the variation of the proportion of HIV/AIDS class for different values of γ, showing that the class decrease with increase in γ, the rate which the infected develop full blown AIDS will drastically reduce.

Figure (5) above shows variation of proportion of treatment class for different values of (w_2), the public health campaign for infectives to receive treatment, the class increase initially, and drop because the population of the infected will drop significantly.

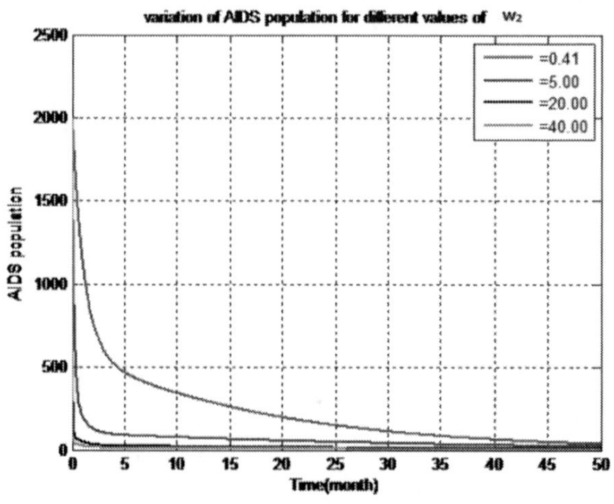

Figure 5. Variation of AIDS population for different values of w_2

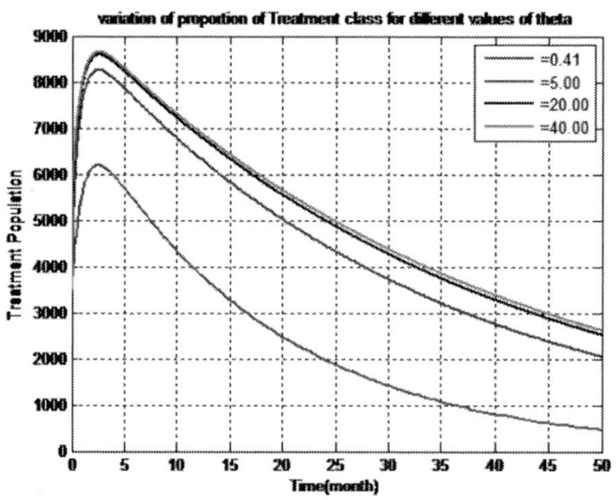

Figure 6. Variation of the population of Treatment class for different values of theta

Figure (6) above shows the variation in the AIDS population for different values of θ. Is is seen that as the θ increases, the AIDS population decrease and increase the treatment rate.

4 Discussion and conclusion

A non-linear mathematical model has been formulated to study the effect of public health campaign on vaccination and treatment for the control of HIV/AIDS within a population. The diseases free equilibrium and endemic equilibrium were obtained and their stabilities investigated. The model showed that the diseases free equilibrium is locally stable when $R_o < 1$ and unstable when $R_o > 1$. The basic reproduction number (R_o) was computed using the next generation method. The endemic equilibrium is locally asymptotically stable using the trace

https://doi.org/10.1051/itmconf/20203103003

and determinant method.
A numerical study of the model was carried out to see the effect of public health campaign on the control of the HIV/AIDS virus. The analysis shows that public health campaign on vaccination has the power to reduce the transmission of the diseases and campaign on therapeutic doses has the power to reduce the diseases burden and the transmission of the dreaded virus; by reducing the stigmatisation and consequently reduced the diseases transmission rate. The public education on treatment in every public gathering for those infected with the virus will make those hiding from the public to present themselves for test and treatment which will result in the decline of AIDS diseases/pandemic.
In conclusion, we recommend that the most effective way to reduce the transmission of HIV/AIDS infection is through public health campaign on vaccination when available and screening for individual to know their HIV status, and if an individual is found positive, the person should be educated on the need to willingly undergo ARV treatment and therapy to reduce the viral load and consequently prolonging life of the individual. Education in public place for safe sex should be enhanced and should be encouraged and other preventive measures such as handling unsterilized needle or needle exchange or any activities that may involved exchange of the bloodily fluid containing blood. This means that the national HIV/AIDS programme in most developing countries should keep on encouraging save sex, advocated free counselling and testing and free ARV treatment, if this is maintained the control of HIV/AIDS will be fully achieved within a short time especially within the sub Saharan Africa. HIV/AIDS is much more pronounce in a very poor and undeveloped nation, therefore more advocacy should be enhanced and more centres should be open to allowed free access to treatment and therapy and which will reduce stigmatization on those living with virus therefore reduce the transmission of HIV virus and control the AIDS epidemic all over the world.

References

[1] A. Kramer, M. Kretzschmar, K. Krickeberg, *Modern Infectious Disease Epidemiology* (Springer, New York, 2010) 443.

[2] W.N. Rom, S.B. Markowitz, *Environmental and Occupational Medicine* (Philadelphia : Wolters Kluwer/Lippincott Williams and Wilkins, 2007), 1884.

[3] World Health Organization, HIV/AIDs Fact Sheet. Retrieved from: https://www.who.int/news-room/fact-sheets/detail/hiv-aids (2019 November 15, accessed 2020 January 29).

[4] L.O. Kallings, Journal of internal medicine **263(3)**, 218–243 (2008).

[5] P.M. Sharp, B.H. Hahn, Cold Spring Harbour Preparative in medicine **1(1)**:a006841 (2011).

[6] G.E. Gray, F. Laher, E. Lazarus, B. Ensoli, L. Corey, Current opinion in Virology **17**, 104–109 (2016).

[7] R.M. Anderson, J. Swinton, G.P. Garnelt, Proceedings. Biological sciences **261(1361)**, 147–151 (1995).

[8] R.M. Anderson, G.P. Garneth Lancet **348(9033)**, 1010–1013 (1996).

[9] D.K. Owens, D.M. Edwards, R.D. Shachter, AIDS **12(9)**, 1057–1066 (1998).

[10] S. Blower, K. Koelle, J. Mills, in *Quantitative Evaluation of HIV prevention programs* E. Kaplan and R. Brookmeyer (Eds.) (Yale University Press, New Haven, CT, 2002) 260–289.

[11] R.H. Gray, X. Li, M.J. Wawer, S.J. Gange, D. Serwadda, N.K. Sewankambo, R. Moore, F. Wabwire-Mangen, T. Lutalo, T.C., Rakai Project Group, AIDS **17(13)**, 1941–1951 (2003).

[12] Z. Mukandavirea, W. Garira, J.M. Tchuencheb, Applied Mathematical Modelling **33(4)** 2084–2095 (2009).

[13] A. Gumel, X. W. Zhang, P. N. Shivakumar, M. L. Garba, B. M. Sahai, Journal of Theoretical Medicine **4(2)**, 147–155 (2002).

ITM Web of Conferences **31**, 03004 (2020) https://doi.org/10.1051/itmconf/20203103004
Mathematical Modelling in Biomedicine 2019

Sensitivity analysis of the model of tuberculosis

Irina Petrova[1,*] and *Dmitry* Gromov[1,**]

[1]Faculty of Applied Mathematics and Control Processes, St.Petersburg State University, Universitetskii prospekt 35, Petergof, Saint Petersburg, Russia.

Abstract. In this contribution, a sufficiently simple, yet realistic model of TB propagation and control is studied in detail. We compute the controlled basic reproduction number $R_0(u_B, u_D)$ for the extended model including the action of two types of controls and analyze the sensitivity of $R_0(u_B, u_D)$ with respect to both controls following the recently devised approach. It is shown that the computed sensitivity coefficients admit a clear epidemiological interpretation and can be used in assessing the efficacy of the respective types of intervention.

1 Introduction

About one quarter of the world's population is infected with tuberculosis (TB) bacteria. Only a small proportion of those infected will become sick with tuberculosis. However, nowadays TB is one of the top 10 causes of death and the leading cause from a single infectious agent worldwide [1]. People fell ill with TB in all countries and all age groups, but individuals with weakened immune system have a much greater risk of getting infected with TB.

Since decades, there have been long-standing efforts to produce an adequate model of population level dynamics of tuberculosis. We mention the seminal papers [2] and [3] as well as an extensive overview [4]. In our work, we deal with the simple model of tuberculosis introduced in [5]. This model has two control parameters describing the rates of detecting infected individuals from different categories.

Recently, a number of results devoted to the structural analysis of compartmental models were obtained, see, e.g., [6, 7]. In particular, the notion of the sensitivity coefficients R_1 was introduced in [6]. The main goal of this study is to apply the results developed in the mentioned paper to the population level dynamical model of tuberculosis and derive conclusions aimed at improving the treatment strategy.

This paper consists of three sections. In Section 2, the structure of the model of tuberculosis epidemiology is considered. In Section 3 we derive the controlled basic reproductive number and perform the sensitivity analysis. Finally, in Section 4 we present numerical results based on real statistical data.

2 The model of tuberculosis

We shall consider a population balance model of tuberculosis. The total population is assumed to be heterogeneous, that is it consists of individuals whose properties or behavior

*e-mail: brutova96@gmail.com
**e-mail: dv.gromov@gmail.com

differ in some respect. The population is split in a number of subgroups (compartments) that are considered to be homogeneous. That is all individuals within a compartment are identical in their evolution.

When infected with TB bacteria, an individual does not necessarily become sick. Thus, we distinguish two TB related conditions: latent TB infection and (active) TB disease. In the former case the immune system keeps the bacteria under control and so, people with latent TB infection are not ill and cannot transmit the disease. People infected with TB bacteria have a 5–15% lifetime risk of developing active form of TB. If, however, the immune system cannot keep the bacteria under control, the bacteria multiply and destroy tissue. The bacteria usually attack the lungs, but can attack any part of the body. At this point, the person has active TB disease.

We further distinguish two types of active TB: open or smear-positive (SP) and close or smear-negative (SN). In the case of open tuberculosis an inflammation develops inside the lungs. This enables TB bacteria to spread by airborne transmission. In the case of closed tuberculosis, there is no risk of infection as the inflammation is on the lungs periphery and not inside.

Taking into account the above classification, the population is split in 6 groups and the respective states correspond to the number of individuals within the groups:

S: number of vaccinated susceptible individuals;

L: number of latently infected individuals without clinical disease manifestations and patients with primary tuberculosis progressing without clear symptoms;

D: number of unidentified active smear-negative individuals;

B: number of unidentified active smear-positive individuals;

D_0: number of identified and treated active smear-negative individuals;

B_0: number of identified and treated active smear-positive individuals.

The dynamics of the system include transitions between subgroups as well as the in- and outflows associated with the respective subgroups. There are two control parameters: u_D and u_B. They describe the rates at which active smear-negative and smear-positive individuals are identified and enrolled in treatment.

The dynamics of disease propagation is described by the following differential equations [5], which are rewritten in order to distinguish between the uncontrolled and the controlled parts:

$$\frac{d}{dt}\begin{bmatrix} L \\ D \\ B \\ D_0 \\ B_0 \\ S \end{bmatrix} = \begin{bmatrix} (1-p)\beta S\,(B+kB_0)+\beta_L D+\beta_{L_0} D_0-\alpha\,(B+kB_0)\,L-(\gamma+\mu)\,L \\ p\beta S\,(B+kB_0)+(\gamma+\alpha\,(B+kB_0))\,L+\beta_D B-(\beta_B+\beta_L+\mu_D)\,D \\ \beta_B D-(\beta_D+\mu_B)\,B \\ \beta_{D_0} B_0-(\beta_{B_0}+\beta_{L_0}+\mu_{D_0})\,D_0 \\ \beta_{B_0} D_0-(\beta_{D_0}+\mu_{B_0})\,B_0 \\ -\beta S\,(B+kB_0)-\mu S+f_S \end{bmatrix} + \begin{bmatrix} 0 & 0 \\ -D & 0 \\ 0 & -B \\ D & 0 \\ 0 & B \\ 0 & 0 \end{bmatrix}\begin{bmatrix} u_D \\ u_B \end{bmatrix} \tag{1}$$

with initial conditions $(S(t_0), L(t_0), D(t_0), B(t_0), D_0(t_0), B_0(t_0)) \in \mathbb{R}^6_+$. Note that we consider the case where there is no influx of individuals from outside into L compartment.

In Table 1, all used parameters are listed and explained. We also show the values of the parameters that were used when performing numerical analysis. These numerical values were obtained from fitting statistical data from Orel Region, Russia to the model (1) and reported in [8].

Table 1. Parameters for compartment L

Parameter	Description	Value	Dimension
μ	the rate constant of natural mortality	1.74×10^{-2}	year^{-1}
μ_D	the rate constant of mortality in the D compartment	0.12	year^{-1}
μ_{D_0}	the rate constant of mortality in the D_0 compartment	5.3×10^{-2}	year^{-1}
μ_B	the rate constant of mortality in the B compartment	0.5	year^{-1}
μ_{B_0}	the rate constant of mortality in the B_0 compartment	0.13	year^{-1}
p	the probability of the fast progression of TB	0.05	dimensionless
k	decrease of contagiousness for people on treatment	0.1	dimensionless
γ	the rate constant of latent TB reactivation	1.14×10^{-3}	year^{-1}
α	the rate constant of super-infection	1.45×10^{-6}	$(\text{person} \cdot \text{year})^{-1}$
β	the rate of infection	7.9×10^{-5}	year^{-1}
β_D	the rate of constant of spontaneous remission in the compartment B	0.5	year^{-1}
β_B	the rate of constant of SN-SP transmission without treatment	0.5	year^{-1}
β_L	the rate of constant of spontaneous recovery in the compartment D	0.23	year^{-1}
β_{D_0}	the rate of transmission to remission during treatment	0.31	year^{-1}
β_{B_0}	the rate of SN-SP transmission under treatment	7.7×10^{-2}	year^{-1}
β_{L_0}	the rate of recovery in the compartment D_0	0.44	year^{-1}
f_S	the rate of influx of uninfected young people	12325	person/year
u_D	the rate of detection of SN individuals	0.156	year^{-1}
u_B	the rate of detection of SP individuals	3.21	year^{-1}

3 Sensitivity analysis

We consider the approach to the system analysis based on the concept of basic reproduction number. The basic reproduction number (R_0) is defined as the number of infections generated by an infected person in a fully susceptible population. It is a key determinant of the dynamics of an infectious disease. R_0 can be interpreted as threshold parameter: if $R_0 < 1$ then the disease can not sustain itself in the population and will eventually die out (see, e.g., [9]). That is, to push a disease towards extinction one has to implement a population-level intervention that reduces R_0 as much as possible.

To determine R_0 we present the system (1) in the form

$$\dot{x} = \Phi(x) = \Phi^I(x) + \Phi^C(x) + \Phi^u(x)u \tag{2}$$

where $\Phi^I(x)$ is the rate at which new infections occur in the population, $\Phi^C(x)$ represents all flows not related to new infections, and $\Phi^u(x)$ is the control matrix. The control matrix describes the controlled flows from the infected compartments to other infected compartments (as treated individuals are considered to be infectious, albeit to a lesser extent).

Next, we define a disease-free equilibrium. Let x^* be an equilibrium state, i.e. $\Phi(x^*) = 0$. Then x^* is said to be a disease-free equilibrium (DFE), denoted x_{DFE}, if x^* corresponds to the point at which no disease exists. Otherwise, x^* is referred to as an endemic infection equilibrium.

We will study the system's dynamics when constant controls $(u_D^*, u_B^*) \in \mathbb{R}_+^2$ are applied. The linearized model (1) at DFE $x = x_{DFE}$ has the following form:

$$\dot{x} = \left(A^I + A^C + B_B u_B^* + B_D u_D^*\right) x, \tag{3}$$

where

$$A^I = D\Phi^I(x)\Big|_{x=x^*} = \begin{pmatrix} A_{11}^I & 0 \\ 0 & 0 \end{pmatrix}, \quad A^C = D\Phi^C(x)\Big|_{x=x^*} = \begin{pmatrix} A_{11}^C & 0 \\ A_{21}^C & A_{22}^C \end{pmatrix},$$
$$B_k = D\Phi_k^{u_k^*}(x)\Big|_{x=x^*} = \begin{pmatrix} B_{11,k} & 0 \\ B_{21,k} & 0 \end{pmatrix}, \quad k = B, D.$$

The matrix A_{11}^I has non-negative elements. A_{11}^C is a matrix with non-positive diagonal elements and non-negative off-diagonal elements. The matrices $B_{11,k}$ have non-positive diagonal and non-negative off-diagonal elements. Furthermore, $B_{11,k}$ are weakly column diagonally dominant. We introduce the following notation: $\mathcal{F} = A_{11}^I$, $\mathcal{V} = -A_{11}^C$, and $\mathcal{W}_k = -B_{11,k}$, $k \in \{B, D\}$.

We are interested in stability of the linearized system (3). The structure matrix of (3) is a block-lower triangular matrix, whose eigenvalues coincide with the eigenvalues of the diagonal blocks. The lower right block is Hurwitz. Thus, the stability of (3) is determined by the eigenvalues of the matrix: $J_u = \mathcal{F} - \mathcal{V} - \mathcal{W}_D u_D - \mathcal{W}_B u_B$. And this matrix is Hurwitz if and only if $\rho\left(\mathcal{F}(\mathcal{V} + \mathcal{W}_B u_B + \mathcal{W}_D u_D)^{-1}\right) < 1$, where ρ denotes the spectral radius.

This spectral radius, parametrized by u_D and u_B, is called the *controlled basic reproduction number*:

$$R_0(u_B, u_D) = \rho\left(\mathcal{F}(\mathcal{V} + \mathcal{W}_B u_B + \mathcal{W}_D u_D)^{-1}\right) \tag{4}$$

Obviously, we have $R_0(0,0) = R_0$, where R_0 is basic reproduction number of the uncontrollable system.

This approach for finding $R_0(u_B, u_D)$ is based on next generation matrix method [10]. However, in many cases it is difficult to compute an explicit expression for the controlled basic reproduction number $R_0(u_B, u_D)$. Furthermore, in applications it is desirable to be able to compare the contribution of each of the used controls. Knowing that, a policy maker can redistribute the investments in order to focus on a more efficient strategy.

Following the approach developed in [6], we represent $R_0(u_B^*, u_D^*)$:

$$R_0(u_B^*, u_B^*) = R_0 + R_1^B u_B^* + R_1^D u_D^* + O(\|u\|^2),$$

where R_1^B and R_1^D are the first order sensitivity coefficients, defined as follows:

$$R_1^k = -y_0^T \mathcal{W}_k \mathcal{V}^{-1} x_0 R_0 / (y_0^T x_0), \quad k = B, D. \tag{5}$$

Here, R_0 is the spectral radius of $\mathcal{F}\mathcal{V}^{-1}$ and x_0, y_0 are the right and left eigenvectors of $\mathcal{F}\mathcal{V}^{-1}$ corresponding to R_0.

We say that a control u_k is efficient if the respective sensitivity coefficient $R_1^k < 0$. Furthermore, the most efficient strategy is the one with the smallest value of parameter R_1^k, because it contributes at most to the decrease in $R_0(u_B, u_D)$.

Using the formulas above, we compute R_0, R_1^D and R_1^B for our system (1). The uncontrolled $R_0(0,0)$:

$$R_0 = \frac{\beta\beta_B\,(\gamma + \mu\,p)\,f_S}{\mu P}, \tag{6}$$

where $P = (\beta_D\beta_L + \beta_B\mu_B + \beta_D\mu_D + \beta_L\mu_B + \mu_B\mu_D)\mu + (\beta_B\mu_B + \beta_D\mu_D + \mu_B\mu_D)\gamma$. Note that P is equal to the determinant of the submatrix of $\mathcal{V}$ that corresponds to the uncontrollable subsystem or, which is the same, the product of the eigenvalues of the respective submatrix. This value constitutes an important structural characteristics of the system.

The expressions for R_1^D and R_1^B are found according to (5). After some algebraic manipulations we write R_1^D as follows:

$$R_1^D = -\left(\frac{P_0}{P} - \frac{\beta_{\rm B0}}{\beta_B}k\right)\frac{\gamma_B R_0}{\gamma_{\rm D0}\gamma_{\rm B0} - \beta_{\rm D0}\beta_{\rm B0}}, \tag{7}$$

where $P_0 = (\mu_{\rm B0}\mu_{\rm D0} + \beta_{\rm D0}\beta_{\rm L0} + \beta_{\rm B0}\mu_{\rm B0} + \beta_{\rm D0}\mu_{\rm D0} + \beta_{\rm L0}\mu_{\rm B0})\mu + (\mu_{\rm B0}\mu_{\rm D0} + \beta_{\rm B0}\mu_{\rm B0} + \beta_{\rm D0}\mu_{\rm D0})\gamma$ and $\gamma_L = \gamma + \mu$, $\gamma_D = \beta_B + \beta_L + \mu_D$, $\gamma_B = \beta_D + \mu_B$, $\gamma_{D0} = \beta_{B0} + \beta_{L0} + \mu_{D0}$, $\gamma_{B0} = \beta_{D0} + \mu_{B0}$ are the total outflow rates from the compartments L, D, B, D_0, and B_0.

Note that the structural parameter P_0 can be interpreted in the same way as P. Namely, it is the product of the eigenvalues of the submatrix corresponding to the states L, D_0, and B_0. The expression (7) can be analyzed to assess the efficacy of the control u_D. Namely, we have that this control is efficient if it holds that $\frac{P_0}{P} > \frac{\beta_{\rm B0}}{\beta_B}k$.

Similarly, the expression for R_1^B is computed to be

$$R_1^B = -\left[\frac{\gamma_D}{\gamma_B}\left(\frac{P_0}{P} - \frac{\gamma_{\rm D0}}{\gamma_D}k\right) + \frac{\gamma\beta_L\mu_D\gamma_{\rm B0}}{\gamma_B P}\left(\frac{\beta_{\rm L0}}{\beta_L} - \frac{\mu_{D0}}{\mu_D}\right) + \frac{\gamma\mu_{\rm B0}\beta_L\beta_B}{\gamma_B P}\left(\frac{\beta_{\rm L0}}{\beta_L} - \frac{\beta_{B0}}{\beta_B}\right)\right]\frac{\gamma_B R_0}{\gamma_{\rm D0}\gamma_{\rm B0} - \beta_{\rm D0}\beta_{\rm B0}} \tag{8}$$

This expression has a component similar to that in (7), but also two additional ones proportional to the differences of ratios of parameters related to the treated and untreated compartments.

4 Numerical computation

To illustrate the theoretical results we first calculated the value of the uncontrolled basic reproduction number, i.e., with $u_D = u_B = 0$ according to (4). In doing so we used a particular set of parameter values as in [8] (see Table 1 for the numerical values). The calculations yielded $R_0(0,0) = 5.1776$, which is quite a large value comparable with those of Polio, Mumps and Smallpox. This stresses the importance of choosing appropriate measures to tackle the disease.

At the next step we calculated the controlled basic reproduction number $R_0(u_B, u_D)$ for the values of controls taken from the interval $[0, 3.5]$. The resulting plot is presented in Fig. 1. The value of the basic reproduction number corresponding to the estimated values of controls $u_D^* = 0.2$ and $u_B^* = 3.1$ is equal to $R_0(u_B^*, u_D^*) = 1.4279 > 1$. This value is indicated by the red dot on the plot.

In Figure 1 we see that $R_0(u_D, u_B)$ decreases as the controls grow. The red line separate the region when R_0 is larger then 1 and the region where it is less than 1.

Using the expressions (7) and (8) we computed numerical values for sensitivity coefficients: $R_1^D = -8.0227$ and $R_1^B = -5.9232$. We can observe that $|R_1^D| > |R_1^B|$, which implies that

 https://doi.org/10.1051/itmconf/20203103004

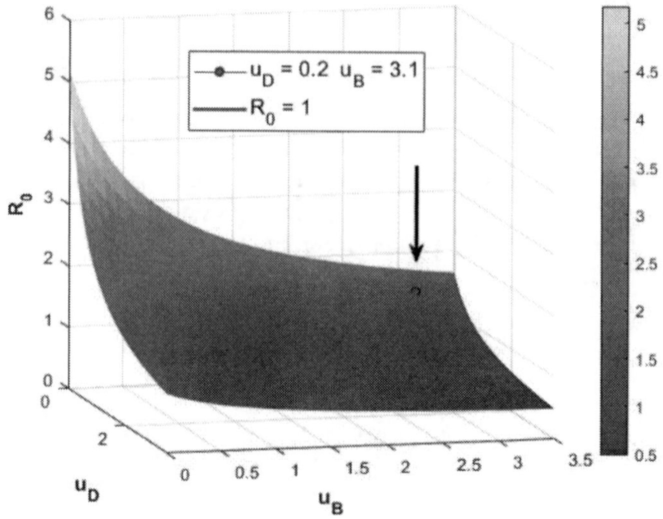

Figure 1. Plot of the controlled basic reproduction number $R_0(u_B, u_D)$ computed for different values of controls.

one can achieve the required reduction in R_0 by screening more sputum-negative individuals. Obviously, screening SN individuals requires larger investment. This leads to the formulation of optimal investment problem that is studied in an ongoing project and will be reported elsewhere.

References

[1] WHO, Tech. rep., World Health Organization (2018), `https://apps.who.int/iris/handle/10665/274453`

[2] M. Perelman, G. Marchuk, S. Borisov, B. Kazennykh, K. Avilov, A. Karkach, A. Romanyukha, Russian Journal of Numerical Analysis and Mathematical Modelling **19**, 305 (2004)

[3] C. Castillo-Chavez, B. Song, Mathematical biosciences and engineering **1**, 361 (2004)

[4] K.K. Avilov, A.A. Romanyukha, Automation and Remote Control **68**, 1604 (2007)

[5] A. Melnichenko, A. Romanyukha, Mathematical Models and Computer Simulations **1**, 428 (2009)

[6] D. Gromov, I. Bulla, E.O. Romero-Severson, Journal of Theoretical Biology **462**, 381 (2019)

[7] D. Gromov, *Stuctural analysis of a controlled compartmental model of disease propagation*, in *Proc. 2019 SICE International Symposium on Control Systems (SICE ISCS)* (2019)

[8] A.A. Romanyukha, *Mathematical models in immunology and epidemiology of infectious deceases* (Binom, Moscow, 2012), ISBN 978-5-9963-0980-1

[9] R.M. Anderson, R.M. May, *Infectious diseases of humans: dynamics and control* (Oxford university press, 1992)

[10] P. Van den Driessche, J. Watmough, Mathematical biosciences **180**, 29 (2002)

Brownian dynamics simulation of cytochrome c diffusion and binding with cytochrome c1 in mitochondrial crista

Anna M. Abaturova[1,*], *Nadezda A.* Brazhe[1], *Ilya B.* Kovalenko[1], *Galina Yu* Riznichenko[1], and *Andrei B.* Rubin[1]

[1]Moscow State University, Moscow, 119991 Russia

Abstract. Cytochrome c (Cc) protein shuttles electrons from respiratory chain complex III — from cytochrome c1 (Cc1) subunit — to complex IV during oxidative phosphorylation, in intermembrane space of mitochondria and cristae lumen. With Leigh syndrome (LS), the crista lumen width (CLW) increases, and ATP production declines. One of the questions raised by this situation is to find out how ATP production impairs at LS. Using the simulation of Brownian dynamics, we tested whether the increase in CLW declines respiration at the stage of electron transport of Cc to Cc1. We designed a Brownian dynamics model of horse Cc diffusion and binding with bovine Cc1 in solution by the ProKSim software. The values of the model parameters were estimated to obtain the same dependence of the second-order association rate constant on the ionic strength as in the experiment [1]. Estimated values of the model parameters were used in the model of the reaction in the cristae lumen. The model scene was a parallelepiped. The distance between the two surfaces simulated crystal membranes varied. We received increasing of half-life time of Cc diffusion and binding with Cc1 at increasing CLW. For membrane surface 90Åx100Å (close to the membrane size of complex III), the half-life time of the process changed from 0.098 to 0.22 μs with increasing cristae lumen width from 120 to 160 Å. But due to the half-life time of electron transfer between proteins in the complex, estimated in [1], is higher (100.5μs), the overall time shouldn't change. To simulate impair of ATP production in the model with an increase in the crista lumen width, we probably need to add to the model IV complex and take into account the dimerization defect of ATP synthase.

1 Introduction

For most cell types in the human body, mitochondria are the central sources of ATP production. In aerobic respiration, membrane-bound protein complexes pass electrons from NADH or succinate to oxygen, pumping protons across the membrane, and creating a proton gradient used by ATP synthase. The respiratory chain complexes and ATP synthase together form the oxidative phosphorylation (OXPHOS) system – figure 1. Respiratory chain complexes consist of four multi-subunit complexes named complex I – complex IV. The electron transfer from NADH to O_2 releases a lot of energy, which is trapped by the I, III, and IV complexes in transporting protons across the inner mitochondrial membrane for ATP synthase to synthesize ATP. Oxygenic phosphorylation entails the transfer of electrons from NADH to

*Corresponding author: e-mail: abaturova@list.ru

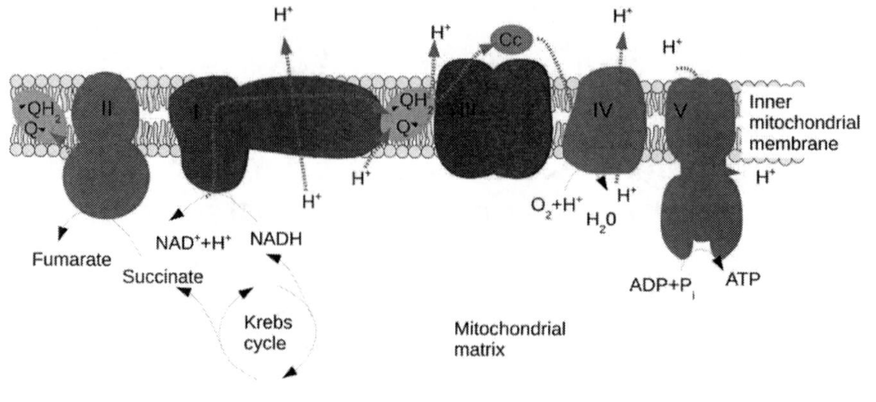

Figure 1. Scheme of oxidative phosphorylation. I – complex I (NADH dehydrogenase), II – complex I (succinate dehydrogenase), Cc – cytochrome c, III – complex III (cytochrome bc1 complex), IV – complex IV (cytochrome c oxidase), V – complex V (ATP synthase), Q – ubiquinone, QH2 – ubiquinool

molecular oxygen. In the respiratory chain, electrons pass from complex I and II to complex III via the hydrophobic electron carrier ubiquinol and from complex III (from subunit cytochrome c1) to complex IV via cytochrome c (Cc). Cc is a 12-kDa soluble protein, diffusing in intermembrane space of mitochondria and cristae lumen. The mitochondrial respiratory chain complexes reside in the inner mitochondrial membrane or cristae. Cristae are the folds of the inner mitochondrial membrane. The respiratory chain complexes form supercomplexes (SC) in the inner mitochondrial membrane [2, 3].

Many studies refer to an altered mitochondrial ultrastructure as an indicator of mitochondria dysfunction. Changes in cristae number and shape define the respiratory capacity as well as cell viability [4]. Leigh syndrome (LS) is a mitochondrial disease affecting mitochondrial energy production via OXPHOS. It caused by several mutations and is a progressive impairment of cognitive and motor functions and premature mortality [5]. In [6] using cryoelectron tomography, were found deep ultrastructural defects in mitochondria of patients, including bloated balloon-like cristae. Crista lumen width CLW (the distance between the two opposite sides of the cristae lumen) was significantly increased: 164±59 Å in patient versus 120±32 Å in healthy control.

2 Description of the model

We designed a Brownian dynamics computer model of diffusion and binding of oxidized horse Cc molecules with water-soluble part of reduced bovine cytochrome c1 (Cc1) in solution by the ProKSim software [7]. We used structures of Cc1 with PDB ID 1BGY [8] and Cc with PDB ID 3O1Y [9]. In the model, when two molecules approached each other, they were directed by the electrostatic field and could occupy a favorable position for docking. In the right orientation, proteins form the so-called transient, or encounter, complex [10]. Further conformational changes in protein molecules result in the formation of the final complex, in which the reaction of electron transfer takes place. In our model, we consider only electrostatic interactions and do not explicitly simulate the final complex formation and electron transfer. We assume that the transient complex transforms into the final one faster than the transient complex forms.

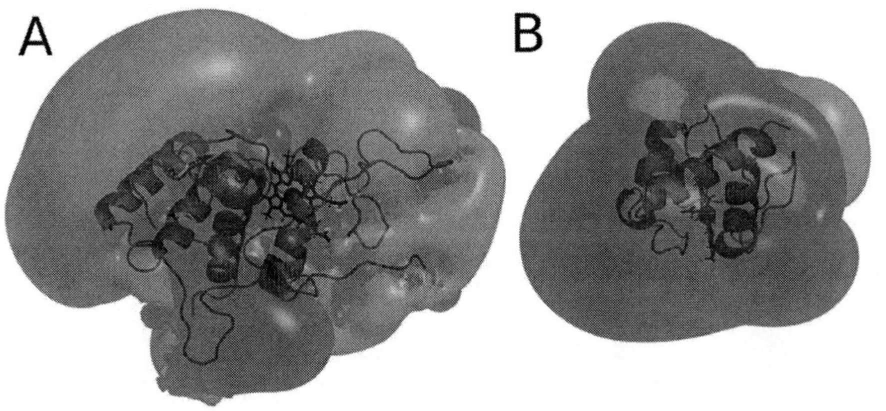

Figure 2. Equipotential surfaces of Cc1 (A) and Cc (B) at I=130 mM. Dark grey (blue) corresponds to potential + 6.5 mV, light gray (red) to - 6.5 mV. Heme is shown by sticks, the Fe atom - as a sphere.

2.1 Electrostatic potentials of proteins

Charged amino acid residues and partial charges of the proteins produce a heterogeneous electrostatic field. If a protein is far away from other proteins, it moves under the effect of the Brownian force. When two protein molecule comes close to each other, their behavior is directed by the electric field of both particles, so that a favorable position for the formation of a transient complex can be achieved. The Poisson-Boltzmann formalism was used to determine the electrostatic potential field generated around the proteins. The step of the grid for calculation potential was 1 Å, the dielectric constant of the cells inside the protein was assigned 2; inside the solution — 80; at the boundary protein solution — 40; and the value of the ionic strength inside the protein was assigned 0. The electrostatic charge distributions on protein atoms are according to the CHARMM27 field [11] and those on the heme, according to [12]. pH in the model was 7. Protein potentials and the molecules were visualized in PyMol (http://pymol.org/).

Figure 2 presents equipotential surfaces of Cc, Cc1 at pH =7 and I=130 mM. Dark grey corresponds to positively charged regions, light grey to negatively charged ones. The Cc1 molecule is negatively charged (net charge -14) and its part, facing the cristae space in mitochondria and binding Cc molecule, has a region of negative potential (figure 2A). On its other part, the molecule has small regions of positive potential. The Cc molecule is positively charged (net charge +8), and its cofactor heme is located in the region of positive potential (figure 2B). The molecule has two negatively charged regions. This inhomogeneity of the equipotential surface should facilitate the proper orientation of Cc relative to its protein partners. The reaction rate decreases with increasing ionic strength, indicating that the interaction has an important electrostatic component [1].

2.2 The model scene in solution

We estimated values of the model parameters (the docking distance and the docking energy) to get the same dependence of the association second-order rate constant on the ionic strength as in experiment at pH 7 [1]. If at some point in time the distance between Fe-atoms of hemes in cytochromes was less than the docking distance, and the electrostatic energy of

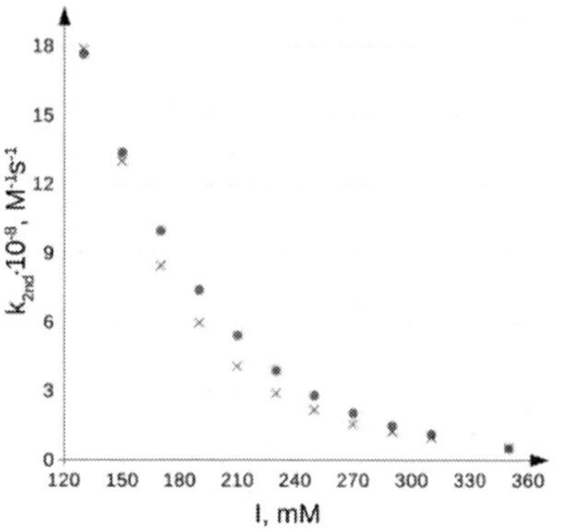

Figure 3. Experimental (circles) and model (cross) dependence of the association rate constant of Cc–Cc1 complex formation on ionic strength.

complex was less than the docking energy, the model molecules formed a transient complex. We chose the docking distance equaled to 35 Å, and the docking energy equaled to -3.4kT. Experimental and model dependence of the association rate constant for Cc-Cc1 complex formation on ionic strength is presented in figure 3. At these values of model parameters and I=130 mM the association rate constant for cytochromes was evaluated as $(1.77 \pm 0.03) * 10^9$ $M^{-1}s^{-1}$. In the experiment it was $1.74 * 10^9$ $M^{-1}s^{-1}$ [1].

The paper [1] shows values for the second-order reaction between bovine III complex and horse photooxidized Ru labeled Cc above 130 mM ionic strength since at lower values of ionic strength kinetic curves are determined by intracomplex electron transport. In computer experiments in cristae we took I=130 mM for approximation the reaction at physiological ionic strength (i.e., I =100-150 mM [13]).

2.3 The model scene in cristae

To study kinetical characteristics of Cc and Cc1 complex formation in crista lumen, we used the parameter values estimated for the buffer solution. The model scene was a parallelepiped with soft boundary conditions. CLW — the distance between 2 surfaces, simulated crista membranes, varied. Cc1 molecules were fixed at their initial positions near the center of one crista membrane — figure 4.

Cc molecules diffused under the action of the random Brownian force and electrostatic force and formed a transient complex with Cc1. We repeated simulation 20000 times. We calculated the number of complexes formed at each time step, normalized kinetic curves obtained are shown in figure 5A. From kinetic curves, we estimated the half-life time of the process $t_{1/2}$ — figure 5B.

3 Results

In one part of model experiments in the cristae, the size of the cristae membranes had a fixed value of 90Åx100Å (close to the membrane size of complex III) and CLW changed from 92 to 300 Å. We got approximately linear increasing of $t_{1/2}$ on CLW - circles in figure 5B. That is because the second-order association rate constant is independent of protein concentration, and half-life time is inversely proportional to the product of the association rate constant by concentration and so proportional to CLW. The half-life time of the process changed from 0.098 μs to 0.22 μs with increasing cristae lumen width from 120 to 160 Å (as at LS [6]). For CLW 300 Å the $t_{1/2}$ was 1.7 μs. The half-life time of electron transfer between proteins estimated in [1] as $\ln 2/(6900\ s^{-1}) = 100.5\mu s$ is higher than that time. That means that the

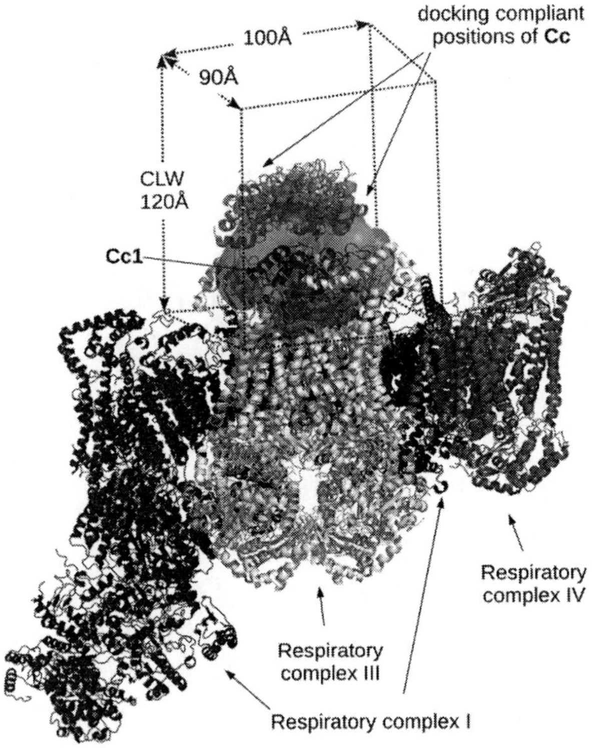

Figure 4. The model scene in cristae. The arrangement of transient Cc–Cc1 complex on the membrane-embedded I1III2IV1 respirasome, obtained by combining Cc1 from PDB ID structure 1BGY and respirasome with PDB ID 5GPN [3]. Shown docking compliant positions of Cc molecules with PDB ID 3O1Y and equipotential surfaces of Cc1 shown as described in figure 2A.

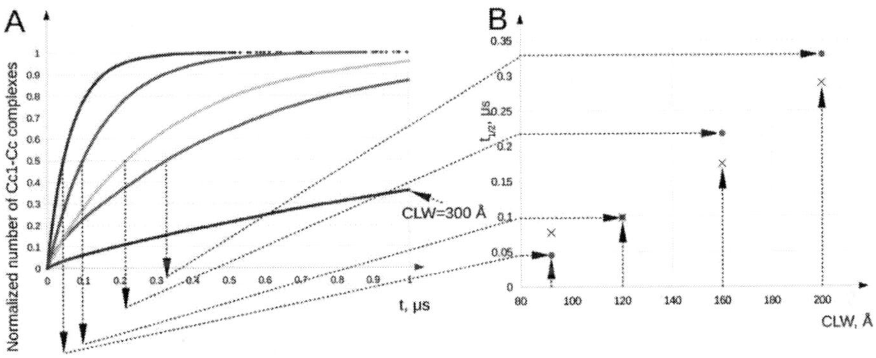

Figure 5. A) Kinetic curves of of Cc1-Cc complex formation for different values of CLW and 90Åx100 Å cristae membranes. B) Dependence of half-life time of Cc diffusion and binding with Cc1 $t_{1/2}$ on CLW; cross - constant value of cristae membranes 90Åx100Å and circles - constant value of volume 90Åx100Åx120Å

electron transfer is the rate-limiting step of Cc-Cc1 interaction. So, the electron transfer rate from Cc1 to Cc, for this size of cristae partition, is determined by the rate of electron transport, and should not depend on CLW. In this case, reaction volume changed linearly on CLW. To exclude the effect of varying concentration at changing CLW, in another part of model experiments, we changed the size of cristae membranes with changing CLW to get the volume as at case with dimensions 90x100x120 Å^3. In this case, the dependence of $t_{1/2}$ on CLW deviates from linear in contrast to the case with the changing volume - cross in figure 5B. In the third part of model experiments, the surface size of the cristae membranes was constant and bigger than dimensions of the respirasome (190Åx300Å [3]) – 900Åx1000Å. In the last case, the half-life time of Cc diffusion and binding with Cc1 increased from 42 μs to 56 μs with increasing cristae lumen width from 120 to 160 Å. So, the reaction became non-limited by electron transport but less sensitive to CLW.

4 Discussion and conclusions

On the Brownian dynamics model of Cc diffusion and binding with Cc1 in the cristae partition of complex III size, we didn't get the impairment of the overall electron transport in the respiratory chain with the increase in CLW. The impair of ATP production may be caused by: 1) the dearth of cristae [1]; 2) the disruption of ATP synthase dimerization [1]; 3) the defective assembly of respiratory chain complexes in SC that may decrease the respiratory chain efficiency [14]. In future studies for the simulation of the drop of ATP production under the condition of the increased crista lumen width, we will add to the model complex IV and the defect of ATP-synthase dimerization.

5 Acknowledgments

The research is carried out using the equipment of the shared research facilities of HPC computing resources at Lomonosov Moscow State University. The work was supported by the Russian Foundation for Basic Research (projects no. 19-04-00999).

References

[1] G. Engstrom, R. Rajagukguk, A. J. Saunders, C. N. Patel, S. Rajagukguk, T. Merbitz-Zahradnik, K. Xiao, G. J. Pielak, B. Trumpower, C. A. Yu, L. Yu, B. Durham, F. Millett, Biochemistry **42**, 2816–24 (2003)

[2] T.J. Jeon, H.M. Kim, S.E. Ryu, Applied Microscopy **48**, 81–86 (2018)

[3] J. Gu, M. Wu, R. Guo, K. Yan, J. Lei, N. Gao, M. Yang, Nature **537**, 639–43 (2016)

[4] R. Quintana-Cabrera, A. Mehrotra, G. Rigoni, M.E. Soriano, Biochem. Biophys. Res. Commun. **500**, 94–101 (2018)

[5] M. Gerards, S.C.E.H. Sallevelt, H.J.M. Smeets, Mol. Genet. Metab. **117**, 300–12 (2016)

[6] S.E. Siegmund, R. Grassucci, S.D. Carter, E. Barca, Z.J. Farino, M. Juanola-Falgarona, P. Zhang, K. Tanji, M. Hirano, E. A. Schon, J. Frank, Z. Freyberg, iScience **6**, 83–91 (2018)

[7] S. Khruschev, A. Abaturova, A. Diakonova, D. Ustinin, D. Zlenko, V. Fedorov, I. Kovalenko, G. Riznichenko, A. Rubin, Computer Research and Modeling **5**, 47–64 (2013)

[8] S. Iwata, J.W. Lee, K. Okada, J.K. Lee, M. Iwata, B. Rasmussen, T.A. Link, S. Ramaswamy, B.K. Jap, Science **281**, 64–71 (1998)

[9] M. De March, N. Demitri, R. De Zorzi, A. Casini, C. Gabbiani, A. Guerri, L. Messori, S. Geremia, J. Inorg. Biochem. **135**, 58–67 (2014)

[10] M. Ubbink, FEBS Lett. **583**, 1060–6 (2009)

[11] A.D. Mackerell, M. Feig, C.L. Brooks, J. Comput. Chem. **25**, 1400–15 (2004)

[12] F. Autenrieth, E. Tajkhorshid, J. Baudry, Z. Luthey-Schulten, J. Comput. Chem. **25**, 1613–22 (2004)

[13] J.D. Cortese, A.L. Voglino, C.R. Hackenbrock, Biochim. Biophys. Acta **1100**, 189–97 (1992)

[14] S. Cogliati, C. Frezza, M.E. Soriano, T. Varanita, R. Quintana-Cabrera, M. Corrado, S. Cipolat, V. Costa, A. Casarin, L.C. Gomes, E. Perales-Clemente, L. Salviati, P. Fernandez-Silva, J.A. Enriquez, L. Scorrano, Cell **155**, 160–71 (2013)

ITM Web of Conferences **31**, 04002 (2020) https://doi.org/10.1051/itmconf/20203104002
Mathematical Modelling in Biomedicine 2019

A model of chronic inflammation in atherosclerosis

Vladimir Bezyaev[1,*], *Nail* Sadekov[1,**], and *Vitaly* Volpert[1,2,***]

[1]RUDN University, 6 Miklukho-Maklaya St, Moscow, 117198, Russian Federation
[2]Institut Camille Jordan, UMR 5208 CNRS, University Lyon 1, Villeurbanne 69622, France

Abstract. Atherosclerosis is a chronic progressive disease leading to the formation of atherosclerotic plaques in artery walls, narrowing its lumen. In this work, we present a mathematical model of atherosclerosis development for the concentrations of cells and cytokines of innate immunity. This system describes a chronic inflammatory response in the intima of an artery vessel wall. The stationary points and their stability for reduced kinetic systems of two and five ODEs are investigated. The relationship of the obtained results is analyzed, and their biological interpretation is given.

1 Introduction

Cardiovascular diseases related to atherosclerosis represent one of the major causes of morbidity and mortality worldwide. Although some progress has been made in diagnosis, surgical treatment and control of high cholesterol level, a decisive progress to fight atherosclerosis has not yet been achieved. Coronary heart disease and stroke result from a sharp decrease in blood supply due to atherosclerotic plaque or related thrombosis and vessel occlusion [1].

The process of plaque development can begin with the damage of the endothelial layer, allowing low density lipoproteins (LDL) to pass from the blood into the intima with further oxidizing by free radicals [2, 3]. Sensing the presence of the oxidized LDL (ox-LDL) in the intima, endothelial cells secrete monocyte chemoattractant protein (MCP-1), which promotes the recruitment of monocytes in the intima [4]. After entering the intima, monocytes differentiate into macrophages, which endocytose the ox-LDL [5–7]. A large amount of ox-LDL in the vessel wall promotes the transformation of macrophages into foam cells [2, 3]. Macrophages M_1 produce pro-inflammatory cytokines such as IL-6, IL-12 and TNF-α and the alternatively activated macrophages M_2 produce anti-inflammatory cytokines such as IL-10 and TGF-β [6]. Pro-inflammatory cytokines contribute to plaque growth activating T_1-helper lymphocytes [8] that produce pro-inflammatory cytokines such as IFN-γ and IL-1β and stimulating M_1 macrophages differentiation, whereas T_2-helper cells produce cytokines such as IL-4 and IL-13 required for M_2 macrophages differentiation [6]. Together with LDL, high density lipoproteins (HDL) also enter the intima. HDL acts to prevent the development of atherosclerosis taking free radicals from oxidizing LDL and possibly removing cholesterol from foam cells [9].

*e-mail: vbezyaev@mail.ru
**e-mail: nail.sadd@mail.ru
***e-mail: volpert@math.univ-lyon1.fr

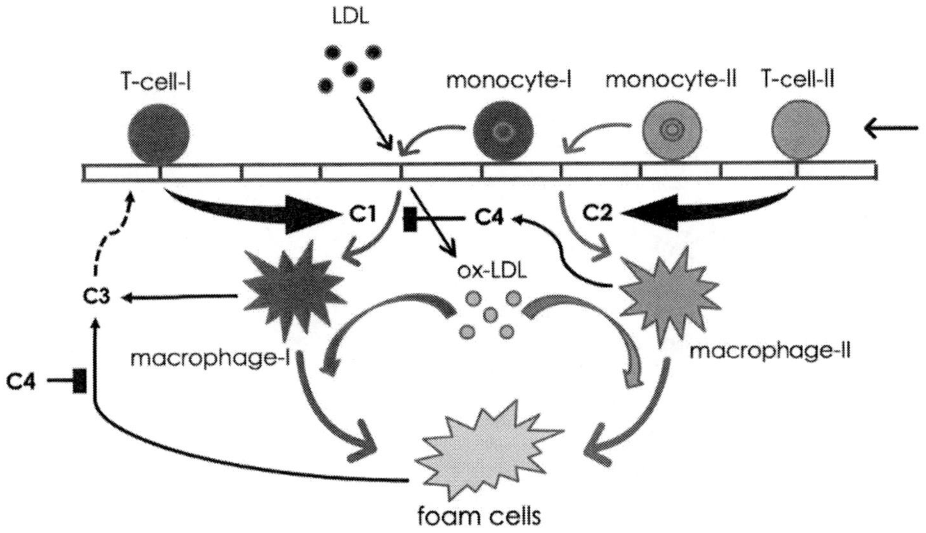

Figure 1. Schematic network of atherosclerosis. LDL in the intima are oxidized and converted to ox-LDL. Ox-LDL activate endothelial cells leading to the recruitment of monocytes. By ingesting ox-LDL, macrophages are transformed into foam cells. Cytokines C_1 and C_2 secreted by T-cells enhance the activity of macrophages. Macrophages M_1 and foam cells F secrete pro-inflammatory agents C_3, which activates T-cells. Macrophages M_2 secrete anti-inflammatory agents C_4, which inhibit T-cells activation.

2 Mathematical model

In this work, we present a mathematical model of atherosclerosis development based on a ODE system for the concentrations of cells and cytokines of innate immunity. This system describes a chronic inflammatory reaction in the intima of the artery wall. The mathematical model is based on the processes schematically shown in Figure 1. The model includes the variables listed in Table 1 and the following equations:

Equation for **LDL** (low density lipoproteins):

$$\frac{dL}{dt} = \sigma_L(L_0 - L) - k_L r L. \tag{1}$$

Here σ_L is rate of LDL penetration through the vessel wall, L_0 is the concentration of LDL in the blood, k_L is the reaction rate of oxidization, where LDL is consumed [2, 3].

Equation for **HDL** (high density lipoproteins):

$$\frac{dH}{dt} = \sigma_H(H_0 - H) - k_H r H. \tag{2}$$

Here σ_H is rate of HDL penetration through the vessel wall, H_0 is the concentration of HDL in the blood, k_H is the reaction rate of oxidization. HDL comes from the blood and it is consumed in the oxidation reaction with free radicals [2, 3].

Equation for **ox-LDL** (oxidized low density lipoproteins):

$$\frac{dL_{ox}}{dt} = k_L r L - \lambda_{L_{ox}M_1}\frac{L_{ox}}{k_{L_{ox}} + L_{ox}} M_1 - \lambda_{L_{ox}M_2}\frac{L_{ox}}{k_{L_{ox}} + L_{ox}} M_2 - d_{L_{ox}} L_{ox}. \tag{3}$$

Table 1. The variables of the model

Variable	Description $[g/cm^3]$
L	Concentration of LDL
H	Concentration of HDL
L_{ox}	Concentration of ox-LDL
r	Concentration of free radicals
A_1	Density of A_1 monocytes
A_2	Density of A_2 monocytes
M_1	Density of M_1 macrophages
M_2	Density of M_2 macrophages
T_1	Density of T_1-helper cells
T_2	Density of T_2-helper cells
C_1	Concentration of T_1-helper cytokines IFN-γ, IL-1β
C_2	Concentration of T_2-helper cytokines IL-4, IL-13
C_3	Concentration of pro-inflammatory agents IL-6, IL-12 and TNF-α
C_4	Concentration of anti-inflammatory agents IL-10 and TGF-β
F	Density of foam cells

Here $\lambda_{L_{ox}M_1}$ and $\lambda_{L_{ox}M_2}$ are the rates of ox-LDL ingestion by M_1 and M_2 macrophages, $k_{L_{ox}}$ is ox-LDL saturation for production of MCP-1, $d_{L_{ox}}$ is degradation rate of ox-LDL. Equation (3) models the production of ox-LDL due to LDL oxidation by reaction with the radicals and a reduction of ox-LDL through ingestion by M_1 and M_2 macrophages [2].

Equation for **free radicals**:

$$\frac{dr}{dt} = r_0 + \lambda_r M_1 - r(k_L L + k_H H). \tag{4}$$

Here r_0 is influx of free radical into intima from the surrounding tissues, λ_r is radicals production rate by macrophages M_1. Equation (4) models the evolution of radical concentration through a source term r_0 [2, 3], production by macrophages M_1 [1] and reduction in concentration through oxidation reactions [2, 3].

Equations for **monocytes**:

$$\frac{dA_1}{dt} = \lambda_{PA_1}\frac{P}{k_{1P} + k_{2P}P}A_1^0 - \lambda_{C_1A_1}\frac{C_1}{k_{C_1} + C_1 + k_4C_4}A_1 - d_{A_1}A_1, \tag{5}$$

$$\frac{dA_2}{dt} = \lambda_{PA_2}\frac{P}{k_{1P} + k_{2P}P}A_2^0 - \lambda_{C_2A_2}\frac{C_2}{k_{C_2} + C_2}A_2 - d_{A_2}A_2. \tag{6}$$

Here λ_{PA_1} and λ_{PA_2} is rate of A_1 and A_2 monocytes penetration through the vessel wall, A_1^0 and A_2^0 is density of A_1 and A_2 monocytes in blood, d_{A_1} and d_{A_2} is death rate of A_1 and A_2 monocytes. Here $P := C_3 + L_{ox}$ corresponds to the activation of endothelial cells (pro-inflammatory action). Migration of smooth muscle cells depends on the activation of endothelial cells, and the activation of endothelial cells depends on P [4], so the coefficient k_{2P} includes the activation of endothelial cells and the migration of smooth muscle cells. The first term is the influx of monocytes into the intima from the blood [1], the second term is the differentiation of monocytes into macrophages [5–7], and the third term corresponds to the death of monocytes.

Equations for **macrophages**:

$$\frac{dM_1}{dt} = \lambda_{C_1A_1}\frac{C_1}{k_{C_1} + C_1 + k_4C_4}A_1 - \lambda_{L_{ox}M_1}\frac{L_{ox}}{k_{L_{ox}} + L_{ox}}M_1 - d_{M_1}M_1, \tag{7}$$

$$\frac{dM_2}{dt} = \lambda_{C_2A_2} \frac{C_2}{k_{A_2} + C_2} A_2 - \lambda_{L_{ox}M_2} \frac{L_{ox}}{k_{L_{ox}} + L_{ox}} M_2 - d_{M_2} M_2. \quad (8)$$

Here first term is the differentiation of monocytes into macrophages [5–7], the second term describes transformation of macrophages into foam cells [2], and the third term corresponds to the death of macrophages.

Equations for **T-helper cells**:

$$\frac{dT_1}{dt} = \lambda_{PT_1} \frac{P}{k_{1P} + k_{2P}P} T_1^0 - d_{T_1} T_1, \quad (9)$$

$$\frac{dT_2}{dt} = \lambda_{PT_2} \frac{P}{k_{1P} + k_{2P}P} T_2^0 - d_{T_2} T_2. \quad (10)$$

Here λ_{PT_1} and λ_{PT_2} is rate of T_1 and T_2 cells penetration through the vessel wall, T_1^0 and T_2^0 is density of T_1 and T_2 cells in blood, d_{T_1} and d_{T_2} is death rate of T_1 and T_2 cells. Equations (9) and (10) models the influx of T-cells from the blood into the intima (similar to monocytes) [1] and death of T-cells.

Equations for **T-helper cytokines** C_1 (IFN-γ, IL-1β) and C_2 (IL-4, IL-13):

$$\frac{dC_1}{dt} = \lambda_{C_1} T_1 - k_{C_1A_1} C_1 A_1 - d_{C_1} C_1, \quad (11)$$

$$\frac{dC_2}{dt} = \lambda_{C_2} T_2 - k_{C_2A_2} C_2 A_2 - d_{C_2} C_2. \quad (12)$$

Here T-cells produced C_1 and C_2 at a rate of λ_{C_1} and λ_{C_2}, which promotes differentiation A_1 and A_2 monocytes into M_1 and M_2 macrophages at a rate of $k_{C_1A_1}$ and $k_{C_2A_2}$ and consumed (internalized) due to this interaction [6]. The third term corresponds to the degradation of cytokines at a rate of d_{C_1} and d_{C_2}.

Equation for **pro-inflammatory agents** (IL-6, IL-12 and TNF-α):

$$\frac{dC_3}{dt} = \lambda_{C_4M_1} \frac{1}{k_{C_4} + C_4} M_1 + \lambda_{C_4F} \frac{1}{k_{C_4} + C_4} F - d_{C_3} C_3. \quad (13)$$

Here macrophages M_1 and foam cells F produce pro-inflammatory cytokines C_3 at a rate of $\lambda_{C_4M_1}$ and λ_{C_4F} activating endothelian cells [6]. It is promote cells penetration. The third term corresponds to the death of cytokines at a rate of d_{C_3}. Anti-inflammatory cytocines C_4 downregulate C_3 production.

Equation for **anti-inflammatory agents** (IL-10 and TGF-β):

$$\frac{dC_4}{dt} = \lambda_{C_4} M_2 - d_{C_4} C_4. \quad (14)$$

Here M_2 macrophages produce anti-inflammatory cytokines C_4 at a rate of λ_{C_4} [5, 6, 10]. It is prevent cells penetration. The second term corresponds to the degradation of cytokines.

Equation for **foam cells**:

$$\frac{dF}{dt} = \lambda_{L_{ox}M_1} \frac{L_{ox}}{k_{L_{ox}} + L_{ox}} M_1 + \lambda_{L_{ox}M_2} \frac{L_{ox}}{k_{L_{ox}} + L_{ox}} M_2 - d_F F. \quad (15)$$

Here M_1 and M_2 macrophages that have ingested a large amount of ox-LDL become foam cells [2]. The third term corresponds to the death of foam cells.

3 Reduced model

Let us consider the particular case where $H = A_2 = M_2 = T_2 = C_2 = C_4 = 0$, $r = R_0 = const$, $A_1 = A_1^0 = const$. From equation (5), assuming a quasi-stationary approximation and $\lambda_{C_1A_1} = 0$, we express A_1 through P and substitute into equation (7). Furthermore, if C_1 is sufficiently large, the term $C_1/(k_{C_1} + C_1 + k_4C_4)$ can be approximated by a constant. Thus, system (1)-(15) can be reduced to the following system of five equations:

$$\frac{dL}{dt} = \sigma_L(L_0 - L) - k_L R_0 L, \tag{16}$$

$$\frac{dL_{ox}}{dt} = k_L R_0 L - \lambda_1 \frac{L_{ox}}{k_1 + L_{ox}} M_1 - d_1 L_{ox}, \tag{17}$$

$$\frac{dM_1}{dt} = \lambda_2 \frac{C_3 + L_{ox}}{k_2 + C_3 + L_{ox}} - \lambda_1 \frac{L_{ox}}{k_1 + L_{ox}} M_1 - d_2 M_1, \tag{18}$$

$$\frac{dC_3}{dt} = \lambda_3 \frac{C_3}{k_3 + C_3} M_1 - d_3 C_3, \tag{19}$$

$$\frac{dF}{dt} = \lambda_1 \frac{L_{ox}}{k_1 + L_{ox}} M_1 - d_4 F, \tag{20}$$

where $\lambda_1 = \lambda_{L_{ox}M_1}$, $\lambda_2 = k_4\lambda_{C_1A_1}$, $\lambda_3 = k_5\lambda_{C_4M_1}$ and $d_1 = d_{Lox}$, $d_2 = d_{M_1}$, $d_3 = d_{C_3}$, $d_4 = d_F$.

It is easy to see that system (16)-(20) for $\lambda_1 = 0$ reduces to the system of two equations

$$\frac{dM_1}{dt} = \lambda_2 \frac{C_3 + L_{ox}^*}{k_2 + C_3 + L_{ox}^*} - d_2 M_1, \tag{21}$$

$$\frac{dC_3}{dt} = \lambda_3 \frac{C_3}{k_3 + C_3} M_1 - d_3 C_3, \tag{22}$$

where (L^*, L_{ox}^*) is the equilibrium point of system (16)-(17) for $\lambda_1 = 0$. The reduced system (21)-(22) of two equations is similar to the model considered in [11]. Thus, for $\lambda_1 = 0$ we know stationary points and the stability.

Then we analyze the system (16)-(20) for $\lambda_1 > 0$. The analysis shows that for $\lambda_1 > 0$, similar to the case $\lambda_1 = 0$, three cases are possible. Case 1: only one equilibrium point E_0 which is stable. Here

$$E_0 = (L^*, L_{ox}^*, M_1^*, 0, F^*), \quad M_1^* = \frac{\lambda_2}{d_2} \frac{L_{ox}^*}{k_2 + L_{ox}^*}, \quad F^* = \frac{\lambda_1}{d_4} \frac{L_{ox}^*}{k_2 + L_{ox}^*}$$

is the easily determined equilibrium point of system (16)-(20) for $C_3^* = 0$. The existence of other (for $C_3 > 0$) equilibrium points is determined, in fact, by the system (17)-(19). Excluding C_3 from the corresponding system of three algebraic equations, we obtain a system of two equations with respect to L_{ox} and M_1:

$$k_L R_0 L^* - \lambda_1 \frac{L_{ox}}{k_1 + L_{ox}} M_1 - d_1 L_{ox} = 0, \tag{23}$$

$$\lambda_2 \frac{L_{ox} + \frac{\lambda_3}{d_3} M_1 - k_3}{L_{ox} + \frac{\lambda_3}{d_3} M_1 - k_3 + k_2} + d_1 L_{ox} - k_L R_0 L^* - d_2 M_1 = 0. \tag{24}$$

Three cases of solving system (23)-(24) are presented in Figure 2 – 4. In case 1, system (23)-(24) has no positive solutions (Figure 1) and, therefore, system (16)-(20) has only one equilibrium point E_0, which is stable. In case 2, system (23)-(24) has only one positive solution (Figure 2) and system (16)-(20) there are two equilibrium points – E_0 is unstable and E_1 is stable (the monostable case). The projection E_1' of the point E_1 on the plane (L_{ox}, M_1) is shown in Figure 3. In case 3, system (23)-(24) has two positive solution (Figure 3) and system (16)-(20) there are three equilibrium points – E_0 and E_2 are stable and E_1 is unstable (bistable case). The projections E_1' and E_2' of the points E_1 and E_2 on the plane (L_{ox}, M_1) is shown in Figure 4. Red lines in Figure 2 – 4 represent graphs of function $M_1(L_{ox})$, and blue lines represent hyperboles in variables (L_{ox}, M_1) defined by system (23)-(24).

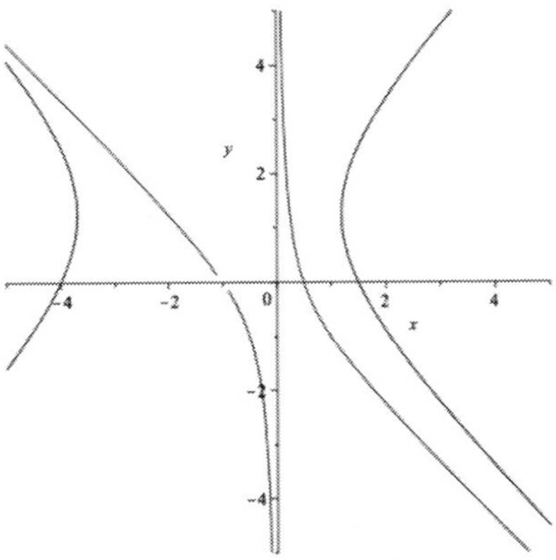

Figure 2. If $\lambda_2\lambda_3 < k_3d_2d_3$ or L_0 small enough and $\lambda_2\lambda_3 > k_3d_2d_3$, system (23)-(24) has no positive solutions.

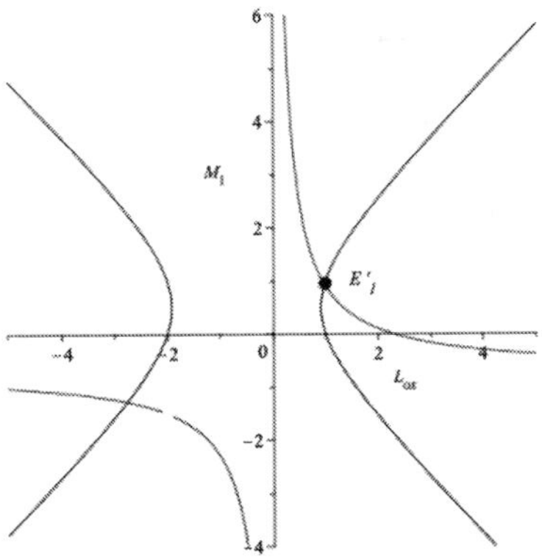

Figure 3. If L_0 large enough and $\lambda_2\lambda_3 > k_3d_2d_3$, system (23)-(24) has only one positive solution.

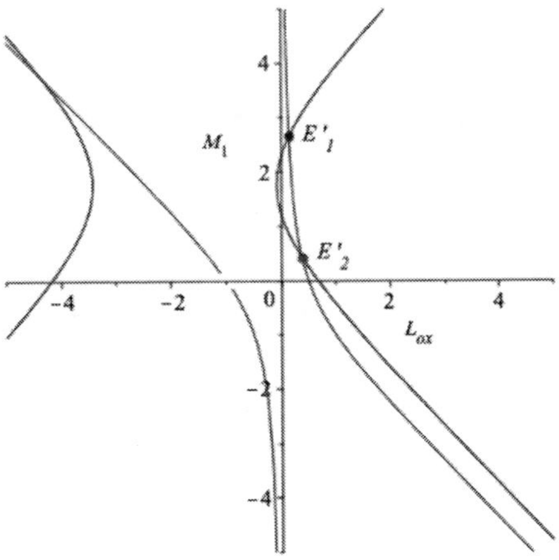

Figure 4. If L_0 is intermediate and $\lambda_2\lambda_3 > k_3d_2d_3$, system (23)-(24) has two positive solutions.

Thus, the analysis of the model of five equations shows that the development of atherosclerosis depends on the concentration L_0 of low-density lipoproteins (bad cholesterol) in the blood. If this concentration is low, then there is only a disease-free equilibrium point, and atherosclerosis does not develop. For some intermediate values, there are three stationary points. Two of them, the disease-free point and the point corresponding to the disease development are stable. Each of them has its own basin of attraction. The system approaches to the second one if the initial perturbation of the disease-free point is sufficiently large. From the physiological point of view, it can be initiated by some additional factors such as inflammation or injury. Finally, for large concentrations of L_0, there are two stationary points. Thee disease-free equilibrium is unstable, and the disease will necessarily develop for any small perturbation.

References

[1] Orekhov A.N., *Atherosclerosis* (Palmarium Academic Publishing, Saarbrucken, Germany, 2013).
[2] Friedman A., Hao W., Bulletin of Mathematical Biology **77**, Issue 5, 758–781 (2015).
[3] Hao W., Friedman A., PLoS ONE **9(3)**, 1–15 (2014).
[4] Harrington J.R., Stem Cells **18**, 65–66 (2000).
[5] Gui T, Shimokado A, Sun Y, Akasaka T, Muragaki Y., Mediators Inflamm **2012**, 1–14 (2012).
[6] Dipak P. Ramji, Thomas S. Davies, Cytokine and Growth Factor Reviews **26(6)**, 673–685 (2015).
[7] Gunter Almer, Daniela Frascione and et., Mol. Pharmaceutics **10**, 175–186 (2013).
[8] Hansson G.K., Holm J., Jonasson L., Am. J Pathol. **135**, 169–175 (1989).
[9] Barter P., European Heart Journal Supplements **7**, 4–8 (2005).
[10] Xinbing Han, William A. Boisvert, *The Role of IL-10 in Atherosclerosis* (InTech, 2012).
[11] El Khatib N., Genieys S., Volpert V., Math. Model. Nat. Phenom. **2**, No. 2, 126–141 (2007).
[12] Panfilov A.V., Dierckx H., Volpert V., Physica D **399**, 1–34 (2019).

Delay Differential Model for Tumour-Immune System With Chemo-therapy and Optimal Control

Fathalla A. Rihan[1,*] and *Bassel F.* Rihan[2]

[1]Department of Mathematical Sciences, College of Science, Al Ain, 15551, UAE University, UAE
[2]Faculty of Medicine, Ain Shams University, Abbassyia 1156, Cairo, Egypt

Abstract. In this paper, we present a mathematical model of tumour-immune interactions in presence of chemotherapy treatment. The model is governed by a system of delay differential equations with optimal control variables. The control variables are included to justify the best strategy of treatments with minimum side effects, by reducing the production of new tumour cells and keeping the number of normal cells above the average of its carrying capacity. Existence of optimality and optimality conditions are also proved. The numerical simulations show that the optimal treatment strategy reduces the load of tumour cells and increases the effector cells after few days of therapy.

Keywords: Chemotherapy; DDEs; Hamiltonian; Immune system; Mathematical modeling; Optimal control; Time-lags

1 Introduction

Mathematical modeling of tumour-immune interactions is very complex and has a long history e.g. [1–7]. In [7], Kuznetsov *et al.* modeled the interactions of cytotoxic T-lymphocyte (CTL) response and the growth of an immunogenic tumour. In other contributions [8–10], the authors take into account the penetration of the tumour cells by the effector cells, which simultaneously causes the inactivation of effector cells. In [11], the authors consider the effects of time-delay on the two-dimensional system which represents the basic model of the immune response. They study variations of the stability of the fixed points due to time-delay and the possibility for the occurrence of the chaotic solutions. Yafia [12] analyzed an interaction between the proliferating and quiescent cells tumour with a single delay. He showed the occurrence of Hopf bifurcation as the delay crosses some critical value [13, 14]. Recently, Rihan *et. al* [15] proposed a delay differential model with optimal control for tumour-immune system with external treatments.

We also mention here that there are many problems in biosciences (such as epidemics, harvesting, chemostat, treatment of diseases, physiological control, vaccination) which can be addressed within an optimal control framework for systems of DDEs [16–19]. However, the amount of real experience that exists with optimal control problems (OCPs) is still small.

Consider a DDE with an $m-$dimensional control term $u(t)$ of the form

$$y'(t) = f(y(t), y(t-\tau_1), y(t-\tau_2), \ldots, y(t-\tau_d), u(t), t) \tag{1}$$

*e-mail: frihan@uaeu.ac.ae

and a suitable objective functional $\mathcal{J}_0(u)$

$$\text{Minimize } \mathcal{J}_0(u) = \int_0^{t_f} \mathcal{F}(y(t), y(t-\tau_1), y(t-\tau_2), \ldots, y(t-\tau_d), u(t), t)dt, \tag{2}$$

and subject to control constraint $a \leq u(t) \leq b$, and state constant $y(t) \leq c$, where a and b are the lower and upper bounds. The integrand, $\mathcal{F}(:)$ is called the Lagrangian of objective functional which is continuous in $[0, t_f]$. Additional equality or inequality constraint(s) can be imposed in of $\mathcal{J}_0(u)$.

OCPs using differential models were studied in connection with immune responses to infections; See [20]. In [21], delay model with optimal control is used to describe the interactions between HIV, CD4+ T cells, and cell-mediated immune response. Both the treatment and the intracellular delay are incorporated into the model in order to improve therapies to cure HIV infection. The optimal controls represent the efficiency of drug treatment in inhibiting viral production and preventing new infections. A humoral immune response model was considered in the paper [22] on determining optimal intravenous drug delivery in AIDS patients. The objective was to find a control strategy that minimizes the total drug administered subject to the constraint that patient recovers. Recently, the authors in [15], studied an optimal control problem of delay differential model to describe the dynamics of tumour-immune interactions in presence of immuno-chemotherapy. The model includes constant delays in the mitotic phase to justify the time required to stimulate the effector cells and for the effector cells to develop a suitable response to the tumour cells. By applying optimal control theory, the cost associated with the immuno-chemotherapy is minimized to reduce load of of tumour cells.

Herein, we just consider a very simple delayed tumour-immune competition model, without treatments which is presented in [23]. Assume that $E(t)$ represents effector cells population, such as $CD8^{+}T$ cells and $T(t)$ is the tumour cells population. The model takes the form

$$\begin{aligned}\frac{dE(t)}{dt} &= \sigma + \frac{\rho E(t-\tau)T(t-\tau)}{\eta + T(t-\tau)} - \mu E(t-\tau)T(t-\tau) - \delta E(t),\\ \frac{dT(t)}{dt} &= r_2 T(t)(1-\beta T(t)) - nE(t)T(t),\end{aligned} \tag{3}$$

where σ represents the normal rate (not increased by the presence of the tumour) of the flow of adult effector cells into the tumour side (region). The source of the immune cells is considered to be out side of the system, so it is reasonable to assume a constant influx rate σ. μ is fraction of immune cells inactivated in interactions with tumour cells. In the absence of any tumour, the cells will die at a rate δ. The presence of tumour cells stimulates the immune response, represented by the positive nonlinear growth term for the immune cells $\rho E(t-\tau)T(t-\tau)/(\eta + T(t-\tau))$ where ρ and η are positive constants, $\tau \geq 0$ is the time-delay that presents the time needed by the immune system to develop a suitable response after recognizing the tumour cells. The saturation term (Michaelis-Menten form) with the $E(t)$ compartment and logistic term with the $T(t)$ compartment is considered. Of course, the presence of the tumour cells virtually initiates the proliferation of tumour-specific effector cells to reach a saturation level parallel with the increase in the tumour populations. r_2 is the growth rate of tumour cells. As a result of the interaction, the immune effector cells decrease load tumour cells at rate n.

2 Delay Differential Model of Tumour-Immune System With Chemotherapy

To include external chemotherapy in model (3), we should consider extra two variables namely amount of chemotherapy $u(t)$ and normal cells $N(t)$ with control variable $v(t)$. The formulation as an optimal control problem allows us to: (i) investigate the dynamical system of interacting cell populations being affected by the treatments; (ii) Optimize the application of the control such that the quantity of the treatments is optimized; and (iii) Minimize the tumour size at some of end-time; See [15].

We aim to design an efficient treatment protocol, where we employ the tools of optimal control theory. This demonstrates how immunotherapy and chemotherapy might be combined for more effective treatment and to protect the patient from opportunistic infection, as well as fighting the cancer itself. Unlike chemotherapy, immunotherapy does not kill tumour cells directly, but it activates and stimulates the growth of immune cells, most importantly T-Cells, and NK Cells, which are capable of destroying cancer cells directly. Therefore, the main goal of combining immuno-chemotherapy treatment is to eradicate the tumour cells, with the minimum side-effect, while maintaining adequate amounts of healthy tissues. We also assume a homogeneity of the tumour cells. The modified model is

$$\begin{aligned}
\frac{dE(t)}{dt} &= \sigma + \frac{\rho E(t-\tau)T(t-\tau)}{\eta + T(t-\tau)} - \mu E(t-\tau)T(t-\tau) - \delta E(t) - a_1(1-e^{-u(t)})E(t), \\
\frac{dT(t)}{dt} &= r_2 T(t)(1-\beta T(t)) - nE(t)T(t) - c_1 N(t)T(t) - a_2(1-e^{-u})T(t), \\
\frac{dN(t)}{dt} &= r_3 N(t)(1-\beta_2 N(t)) - c_2 T(t)N(t) - a_3(1-e^{-u})N(t), \\
\frac{du(t)}{dt} &= v(t) - d_1 u(t).
\end{aligned} \tag{4}$$

Here, $N(t)$ is the normal or host cells population, and u is the amount of chemotherapy at the tumour site in the host over time. For $\tau = 0$, the model is reduced to ODEs model developed by de Pillis and Radunskaya in [24]. In addition, the reaction of immune cells and tumour cells can result in either the death of tumour cells or the inactivation of the immune cells, resulting in the two competition terms $-\mu E(t-\tau)T(t-\tau)$ and $-nET$. The tumour cells and normal cells are modelled by a logistic growth law, with parameters r_i representing the growth rate of two types of cells: $i = 2$ identifies the parameter associated with tumour, and $i = 3$ identifies the one associated with the normal tissue, β and β_2 are the reciprocal carrying capacities of tumour cells and host cells respectively. In addition, there are two terms representing the competition between tumour and host cells $-c_1NT$ and $-c_2NT$.

We assume that the drug kills all types of cells, but that the killing rate differs for each type of cells, $a_i(1-e^{-u})$ is the fraction cell kill for a given amount of drug, u, at the tumour site. We denote by a_1, a_2, and a_3 the three different response coefficients. $v(t)$ represents the amount of dose that is injected into the system, while d_1 is the decay rate of the drug once it is injected. In this case, the quantity we will control directly is not u, but $v(t)$.

It is easy to show that there exists a unique solution of system (4) with initial data $(E_0, T_0, N_0, u_0) \in C$. For biological reasons, we assume that the initial data for system (4) satisfy

$$E_0 \geq 0, \quad T_0 \geq 0 \quad N_0 \geq 0, \quad u_0 \geq 0.$$

Optimality in treatment might be defined in a variety of ways. Some studies have been done in which the total amount of drug administered is minimized, or the number of tumour cells is minimized. The general goal is to keep the patient healthy while killing the tumour.

Since our model takes into account the toxicity of the drug to all types of cells, our control problem consists of determining the function $v(t)$ that will maximize the amount of effector cells and minimize the number of tumour cells and the cost of the control with the constraint that we do not kill too many normal cells. If the units of cells are normalized, so that the carrying capacity of normal cells is 1, we then require that the number of normal cells stay above three-fourth of the carrying capacity. Therefore, in the language of optimal control theory, we should provide an objective functional

$$\mathcal{J}(v) = \int_0^{t_f} \left(E - T - \frac{B_v}{2}[v(t)]^2\right) dt \tag{5}$$

where B_v is a weight factor that describes the patient's acceptance level of chemotherapy; and a constraint

$$k(E, T, N, u, E_\tau, T_\tau, v) = N - 0.75 \geq 0, \qquad 0 \leq t \leq t_f. \tag{6}$$

We choose the control parameter as a class of piecewise continuous functions defined for all t such that $0 \leq v(t) \leq v_{max} < \infty$, where $v(t) = v_{\max}$ represents the maximum chemotherapy, while $v(t) = 0$ represents the case where there is no chemotherapy. Thus, we depict the class of admissible controls as

$$V = \left\{v(t), \text{ piecewise continuous, } \mid 0 \leq v(t) \leq v_{max} < \infty, \ \forall t \in [0, t_f]\right\}.$$

If we consider the objective functional as a function of v, it can be seen that J is concave. Hence, a maximum value can be obtained. The goal is to characterize the optimal control v^* satisfies

$$\max_{0 \leq v \leq v_{max}} \mathcal{J}(v) = \mathcal{J}(v^*). \tag{7}$$

2.1 Existence of an optimal control

To prove the existence of the optimal control for the underlying model (4), we use the results of Fleming and Rishel [25, Theorem 4.1, p. 68-69]

Theorem 1 *There exists an optimal control $v^* \in V$ for system (4) such that*

$$\mathcal{J}(v^*) = \max_{v \in V} \mathcal{J}(v) \tag{8}$$

if the following conditions are satisfied: (i) The set of controls and corresponding state variables is nonempty; (ii) The control set V is convex and closed; (iii) The right-hand side of the state system is bounded by a linear combination of the state and control variables; (vi)The integrand of the objective functional is concave on V; (v) There exist constants h_2, $h_3 > 0$ and $b > 1$ such that the integrand $L(E, T, v)$ of the objective functional satisfied $L(E, T, v) \leq h_2 - h_3 (|v|)^b$.

Proof. In order to verify the above conditions, we should first prove the existence of the solution for system (4). Since $\frac{\rho T(t-\tau)}{\eta + T(t-\tau)} < \rho$, $v_{max} < \infty$ and by neglecting the negative terms in the model, we have

$$\frac{dE}{dt} < \sigma + \rho E(t - \tau), \qquad \frac{dT}{dt} < r_2 T, \qquad \frac{dN}{dt} < r_3 N, \qquad \frac{du}{dt} < v_{max}. \tag{9}$$

System (9) can be rewritten in the form:

$$\begin{pmatrix} E \\ T \\ N \\ u \end{pmatrix}' < \begin{pmatrix} 0 & 0 & 0 & 0 \\ 0 & r_2 & 0 & 0 \\ 0 & 0 & r_3 & 0 \\ 0 & 0 & 0 & 0 \end{pmatrix} \begin{pmatrix} E(t) \\ T(t) \\ N(t) \\ u(t) \end{pmatrix} + \begin{pmatrix} \rho & 0 & 0 & 0 \\ 0 & 0 & 0 & 0 \\ 0 & 0 & 0 & 0 \\ 0 & 0 & 0 & 0 \end{pmatrix} \begin{pmatrix} E(t-\tau) \\ T(t-\tau) \\ N(t-\tau) \\ u(t-\tau) \end{pmatrix} + \begin{pmatrix} \sigma \\ 0 \\ 0 \\ v_{max} \end{pmatrix}$$

where $' = d/dt$. This system is linear in finite time with bounded coefficients. Then the solutions of this linear system are uniformly bounded. Therefore, the solution of the nonlinear system (4) is bounded and exists [26]. Hence, condition one is satisfied.

Clearly, the second condition is satisfied by the definition of V. System (4) is bilinear in the control variable v and can be rewritten as

$$\vec{F}(t, \vec{X}(t), \vec{X}(t-\tau), v) = \vec{\alpha}(t, \vec{X}) + \vec{\beta}(t, \vec{X}(t-\tau)) + \sigma + v \tag{10}$$

where $\vec{X}(t) = (E, T, N, u)$, $\vec{X}(t-\tau) = (E(t-\tau), T(t-\tau), N(t-\tau), u(t-\tau))$, $\vec{\alpha}$ and $\vec{\beta}$ are the vector valued functions of $\vec{X}(t)$ and $\vec{X}(t-\tau)$ respectively. Using that the solutions are bounded, we have

$$|\vec{F}(t, \vec{X}(t), \vec{X}(t-\tau), v)| \leq |F_1X(t)| + |F_2X(t-\tau)| + |F_3| + |F_4| \leq h_1|\vec{X}| + |\sigma| + |v|$$

where h_1 depends on the coefficients of the system, and

$$F_1 = \begin{pmatrix} 0 & 0 & 0 & 0 \\ 0 & r_2 & 0 & 0 \\ 0 & 0 & r_3 & 0 \\ 0 & 0 & 0 & 0 \end{pmatrix}, \; F_2 = \begin{pmatrix} \rho & 0 & 0 & 0 \\ 0 & 0 & 0 & 0 \\ 0 & 0 & 0 & 0 \\ 0 & 0 & 0 & 0 \end{pmatrix}, \; F_3 = \begin{pmatrix} \sigma \\ 0 \\ 0 \\ 0 \end{pmatrix}, \; F_4 = \begin{pmatrix} 0 \\ 0 \\ 0 \\ v \end{pmatrix}.$$

We also note that the integrand of $J(v)$ is concave in V. Finally,

$$E - T - B_v/2[v(t)]^2 < E - B_v/2[v(t)]^2 \leq h_2 - h_3|v(t)|^2,$$

where h_2 depends on the upper bounds of E and T, and $h_3 = B_v/2$. This completes the proof. □

2.2 Optimality of system (4)

We proved in the above section the existence of optimal control system given by the objective functional (5) for system (4). In order to derive the necessary conditions for an optimal control problem, we use Pontryagin's maximum principle with delay [27]. This principle converts (4), (5), and (8) into a problem of minimizing a Hamiltonian functional, H with

$$H(t, E, T, E_\tau, T_\tau, u, v, \lambda) = E - T - \frac{B_v}{2}[v(t)]^2 + \lambda_1\frac{dE}{dt} + \lambda_2\frac{dT}{dt} + \lambda_3\frac{dN}{dt} + \lambda_4\frac{du}{dt} + \gamma k \tag{11}$$

and $\gamma \geq 0$ with $\gamma(t)k(t) = 0$, where

$$\gamma = \begin{cases} 1 & \text{if } N(t) \leq 0.75, \\ 0 & \text{otherwise.} \end{cases}$$

To minimize the Hamiltonian functional, the Pontryagin's maximum principle [27] is used. Thus, the following theorem is deduced:

Theorem 2 *Given an optimal control v^* and solutions of the corresponding state system (4), there exist adjoint variables λ_i for $i = 1, 2, 3, 4$ satisfying the following*

$$\begin{aligned}
\lambda_1'(t) &= -1 + \lambda_1(t)\left[\delta + a_1(1 - e^{-u^*})\right] + \lambda_2(t)nT^* + \lambda_1(t+\tau)\chi_{[0,t_f-\tau]}\left[\mu T^* - \frac{\rho T^*}{\eta + T^*}\right], \\
\lambda_2'(t) &= 1 + \lambda_2\left[-r_2 + 2r_2\beta T^* + nE^* + c_1N^* + a_2(1 - e^{-u^*})\right] + \lambda_3c_2N^* \\
&\quad + \chi_{[0,t_f-\tau]}\lambda_1(t+\tau)\left[\frac{\rho E^*T^*}{(\eta + T^*)^2} - \frac{\rho E^*}{\eta + T^*} + \mu E^*\right], \\
\lambda_3'(t) &= \lambda_2c_1T^* - \lambda_3\left(r_3 - 2r_3\beta_2N^* - c_2T^* - a_3(1 - e^{-u^*})\right) - \gamma, \\
\lambda_4'(t) &= -\lambda_1(t)a_1e^{-u^*}E^* + \lambda_2(t)a_2e^{-u^*}T^* + \lambda_3(t)a_3e^{-u^*}N^* + \lambda_4(t)d_1,
\end{aligned} \tag{12}$$

with transversality conditions

$$\lambda_i(t_f) = 0, \qquad i = \{1, 2, 3, 4\} \textit{ and } \quad \chi_{[0,t_f-\tau]} = \begin{cases} 1 & \textit{if } t \in [0, t_f - \tau], \\ 0 & \textit{otherwise}. \end{cases}$$

Moreover, the optimal control is

$$v^* = \min\left(v_{max}, \frac{\lambda_4}{B_v}\right). \tag{13}$$

Proof. The adjoint equations and transversality conditions can be obtained by using Pontryagin's maximum principle with delay in state [27] such that

$$\begin{aligned} \lambda_1'(t) &= -\frac{\partial H}{\partial E}(t) - \chi_{[0,t_f-\tau]}(t)\frac{\partial H}{\partial E_\tau}(t+\tau), \quad \lambda_1(t_f) = 0, \\ \lambda_2'(t) &= -\frac{\partial H}{\partial T}(t) - \chi_{[0,t_f-\tau]}(t)\frac{\partial H}{\partial T_\tau}(t+\tau), \quad \lambda_2(t_f) = 0, \\ \lambda_3'(t) &= -\frac{\partial H}{\partial N}, \quad \lambda_3(t_f) = 0, \\ \lambda_4'(t) &= -\frac{\partial H}{\partial u}, \quad \lambda_4(t_f) = 0. \end{aligned} \tag{14}$$

The optimal control v^* can be solved from the optimality condition $\frac{\partial H}{\partial v}(t) = 0$ that is, $-B_v v + \lambda_4 = 0$.

By the bounds in the control set V, it is easy to obtain v^* in the form of (13). The optimality system consists of the state system (4) coupled with the adjoint system (12) with initial and transverality conditions together with the characterization together with the optimal control. Therefore, the optimal system is given as follows:

$$\begin{aligned} \frac{dE^*}{dt} &= \sigma + \frac{\rho E^*(t-\tau)T^*(t-\tau)}{\eta + T^*(t-\tau)} - \mu E^*(t-\tau)T^*(t-\tau) - \delta E^* - a_1(1-e^{-u^*})E^*, \\ \frac{dT^*}{dt} &= r_2 T^*(1-\beta T^*) - \mu E^*(t)T^* - c_1 N^* T^* - a_2(1-e^{-u^*})T^*, \\ \frac{dN^*}{dt} &= r_3 N^*(1-\beta_2 N^*) - c_2 T^* N^* - a_3(1-e^{-u^*})N^*, \\ \frac{du^*}{dt} &= v^* - d_1 u^*, \\ \lambda_1'(t) &= -1 + \lambda_1(t)\left[\delta + a_1(1-e^{-u^*})\right] + \lambda_2(t) n T^* + \lambda_1(t+\tau)\chi_{[0,t_f-\tau]}\left[\mu T^* - \frac{\rho T^*}{\eta + T^*}\right] \\ \lambda_2'(t) &= 1 + \lambda_2\left[-r_2 + 2r_2\beta T^* + nE^* + c_1 N^* + a_2(1-e^{-u^*})\right] + \lambda_3 c_2 N^* \\ &\quad + \chi_{[0,t_f-\tau]}\lambda_1(t+\tau)\left[\frac{\rho E^* T^*}{(\eta + T^*)^2} - \frac{\rho E^*}{\eta + T^*} + \mu E^*\right] \\ \lambda_3'(t) &= \lambda_2 c_1 T^* - \lambda_3\left(r_3 - 2r_3\beta_2 N^* - c_2 T^* - a_3(1-e^{-u^*})\right) - \gamma \\ \lambda_4'(t) &= -\lambda_1(t) a_1 e^{-u^*} E^* + \lambda_2(t) a_2 e^{-u^*} T^* + \lambda_3(t) a_3 e^{-u^*} N^* + \lambda_4(t) d_1, \\ v^* &= \min\left(v_{max}, \frac{\lambda_4}{B_v}\right) \\ E^*(\theta) &= \psi_1(\theta),\ T^*(\theta) = \psi_2(\theta),\ N^*(\theta) = \psi_3(\theta),\ u(\theta) = \psi_4(\theta),\ \theta \in [-\tau, 0], \\ \lambda_i(t_f) &= 0,\ i = \{1, 2, 3, 4\}. \end{aligned} \tag{15}$$

□

The numerical approximations of the optimal control problem are carried out using forward and backward Euler methods. Starting with an initial guess for the value of the controls on the time interval $[0, t_f]$, we solve the state system with control variables (4) using forward Euler' scheme. While, the adjoint system is solved using the solutions of the state system and the transversality conditions backward in time. The Pontryagin-type maximum principle has been derived for retarded OCP, with delays in the state variable when the control system is. subject to a mixed control state, in order to minimize the cost of treatment, reduce the tumour cells load, and keep the number of normal cells above 75% of its carrying capacity.

Figure 1 shows the impact of chemotherapy treatments (with optimal control) when we choose the parameter values in an unstable region ($\sigma = 0.2$, $\rho = 0.2$, and $\tau = 1.5$). The tumour cell population is growing up over time in the absence of chemotherapy, while the presence of treatment helps the immune system to keep the growth of the tumour cells under its control. The Figure shows that the tumour cells can be eradicated at day 20, which in some degree meets the natural facts.

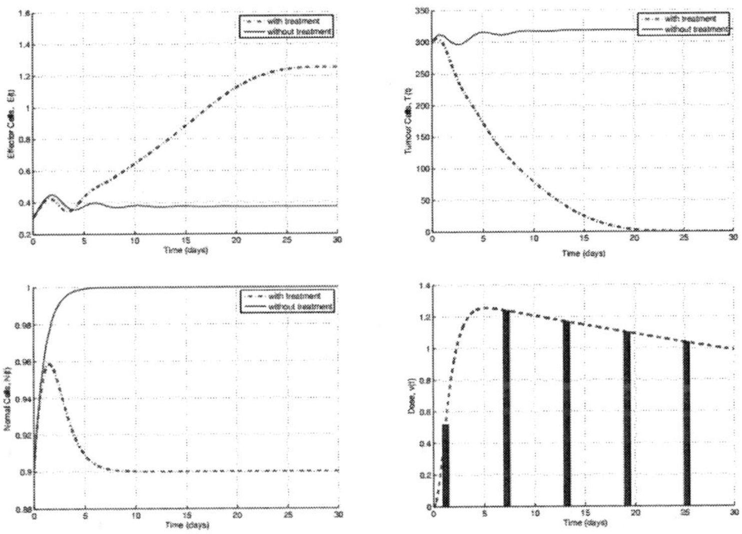

Figure 1. Simulations of system (15), in the stable region, before and after the treatments with control, initial conditions $E_0 = 0.3$, $T_0 = 300$, and $N_0 = 0.9$ and the parameter values are given in the text. The load of tumour cells is growing up over time in absence of treatment; While presence of chemotherapy helps the immune system to keep the growth of the tumour cells under its control.

3 Discussion and Conclusion

Herein, we provided a mathematical model with time-lags and optimal control variables to describe the dynamics of tumour-immune interactions in presence of chemotherapy treatments. The numerical approximations of the optimal control problem are carried out using forward and backward Euler methods. Starting with an initial guess for the value of the controls on the time interval $[0, t_f]$, we solve the state system with control variables (4) using forward Euler' scheme. While, the adjoint system is solved using the solutions of the state system and the transversality conditions (12) backward in time. The Pontryagin-type maximum principle is derived, for retarded optimal control problems with delays in the state variable when the control system is subject to constraints. This is to minimize the cost of treatment, reduce the tumour cells load, and keep the number of normal cells above 75% of its carrying capacity.

The numerical simulations validate the existence of optimality of the control variables. The presence of chemotherapy protocol reduces the tumour load in few days of therapy. Of course, this model can be extended to more powerful models for cancer treatment design with optimal combinations, doses, and scheduling of treatments to speed up the development of individualized therapies.

References

[1] M. Martins, S.C.J. Ferreira, M. Vilela, Phys. Life Rev. **4**, 128 (2007)
[2] R. Araujo, D. McElwain, Bull. Math. Biol. **66**, 1039 (2004)
[3] N. Bellomo, N. Li, P. Maini, Math. Mod. Methods Appl. Sci. **18**, 593 (2008)
[4] N. Bellomo, L. Preziosi, Math. Comput. Model **32**, 413 (2000)
[5] H. Byrne, T. Alarcon, M. Owen, S. Webb, P. Maini, Philos. Trans. R. Soc. A **364**, 1563 (2006)
[6] F. Castiglione, B. Piccoli, J. Theor. Biol. **247**, 723 (2007)
[7] V. Kuznetsov, I.A. Makalkin, M. Taylor, A. Perelson, Bull. Math. Biol. **56**, 295 (1994)
[8] S. Banerjee, Int. J. Appl. Math. Comput. Sci. **18**, 389 (2008)
[9] R. Eftimie, Bull. Math. Biol. **73**, 2 (2011)
[10] F.A. Rihan, S. Muntaser, M.A. Abdeen, D.H. Abdelrahman, J. Appl. Math. **2012**, 19 (2012)
[11] N. Buric, M. Mudrinic, N. Vasovic, Chaos, Solitons and Fractals **12**, 483 (2001)
[12] R. Yafia, Nonlinear Anal. Model. Contr. **11**, 95 (2006)
[13] F.A. Rihan, D.H. Abdelrahman, Int. J. Comput. Math. **90**, 594 (2013)
[14] U. Forys, M. Mochar, J. Polezzczuk, Appl. Math. Letters p. 982986 (2011)
[15] F.A. Rihan, S. Lakshmanan, H. Maurer, Applied Mathematics and Computation **353**, 147 (2019)
[16] H.T. Banks, *Modelling and control in biosciences*, Vol. 6 (Lect. Notes in Biomath: Springer, Berlin, 1975)
[17] V.B. Kolmanovskii, L. Shaikhet, *Control of systems with aftereffect* (American Mathematical Society, 1996)
[18] S.E. Smith, Complexity International **12**, 1 (2005)
[19] F.A. Rihan, B.F. Rihan, Appl. Math. & Inf. Sci. **9**, 1615 (2015)
[20] *Optimal Control Applied to Biological Models* (Chapman and Hall/CRC, New York, 2007)
[21] K. Hattaf, N. Yousfi, ISRN Biomathematics **2012**, 7 pages (2012)
[22] A. Rundell, R. DeCarlo, V. Balakrishnan, H. HogenEsch, IEEE. Trans. Biomed. Eng. **45**, 429 (1998)
[23] F.A. Rihan, D. Abdelrahman, S. Lakshmanan, A. Alkhajah, Appl. Math.Compu. **232**, 606 (2014)
[24] L.G. de Pillis, A. Radunskaya, Comput. Math. Methods Med. **3**, 78 (2001)
[25] W.H. Fleming, R.W. Rishel, *Deterministic and Stochasitic Optimal Control* (Springer-Verlag, New York, 1994)
[26] D.L. Lukes, *Differential Equations: Classical to Controlled* (Academic Press, New York, 1982)
[27] L.S. Pontryagin, R.V. Boltyanski, R.V. Gamkrelidge, E.F. Mischenko, *The Mathematical Theory of Optimal Processes* (John Wiley and Sons, N.Y., 1962)

AUTHOR INDEX

EDP Sciences – Web of Conferences
17, Avenue du Hoggar
Parc d'Activité de Courtabouf
BP 112
F-91944 Les Ulis Cedex A
France

ISBN 978-1-7138-0830-5

CURRAN ASSOCIATES INC.
proceedings.com

Heavy Ion Accelerator Symposium 2015

International Nuclear Structure Conference in
Remembrance of George Dracoulis

Canberra, Australia
14-18 September 2015

edp sciences